WATER RESOURCE PLANNING,
DEVELOPMENT AND MANAGEMENT

THE CRUX OF THE MATTER WATER IN THE REPUBLIC OF THE SUDAN

WATER RESOURCE PLANNING, DEVELOPMENT AND MANAGEMENT

Additional books in this series can be found on Nova's website under the Series tab.

Additional E-books in this series can be found on Nova's website under the E-book tab.

COUNTRIES, REGIONAL STUDIES, TRADING BLOCKS, UNIONS, WORLD ORGANIZATIONS

Additional books in this series can be found on Nova's website under the Series tab.

Additional E-books in this series can be found on Nova's website under the E-book tab.

WATER RESOURCE PLANNING, DEVELOPMENT AND MANAGEMENT

THE CRUX OF THE MATTER WATER IN THE REPUBLIC OF THE SUDAN

ABDEEN MUSTAFA OMER

Nova Science Publishers, Inc.
New York

For permission to use material from this book please contact us:
Telephone 631-231-7269; Fax 631-231-8175
Web Site: http://www.novapublishers.com

LIBRARY OF CONGRESS CATALOGING-IN-PUBLICATION DATA

Omer, Abdeen Mustafa.
The crux of the matter: water in the Republic of the Sudan / Abdeen Mustafa Omer.
p. cm.
Includes index.
ISBN 978-1-61728-752-7 (soft cover)
1. Water-supply--Sudan--Management. 2. Water-supply--Sudan--Statistics. 3. Water quality--Sudan. 4. Water resources development--Sudan. 5. Water--Pollution--Sudan. 6. Sudan--Environmental conditions. 7. Sudan--Social conditions. 8. Sudan--Politics and government--1985- 9. Community life--Sudan. 10. Political participation--Sudan. I. Title.
TD319.S74O46 2010
363.6'109624--dc22

2010025950

Published by Nova Science Publishers, Inc. † New York

CONTENTS

PREFACE

Thirty-nine million people live in Sudan, and environmental pollution is a major concern throughout the country; therefore, industry, communities, local authorities and central government, in order to deal with pollution issues, should adopt an integrated approach. Most polluters pay little or no attention to the control and proper management of polluting effluents. This may reflect lack of enforceable legislation and/or the fear of spending money on the treatment of their effluent prior to discharge. Present resources must be strictly monitored and managed effectively if further deterioration is to be avoided. Community participation in water management programs should fully involve women, who play a central role in the provision and safeguarding of water for domestic use in rural areas. The pricing system recognizes the economic value of water, as stated in the principles of the policy. The harsh climate in Sudan presents unique challenges in meeting growing demands for water and power. The international demand for water increases, compared to the available water resources. In a country with a high population density, there are extreme pressures on water and waste systems, which can stunt the country's economic growth. However, Sudan has recognized the potential to alleviate some of these problems by promoting renewable water and utilizing its vast and diverse climate, landscape and resources, and by coupling its solutions for waste disposal with its solutions for water production. Thus, Sudan may stand at the forefront of the global renewable water community and presents an example of how non-conventional water strategies may be implemented. The present problems that are related to water and sanitation in Sudan are many and varied, and the disparity between water supply and demand is growing with time due to the rapid population growth and aridity. The situation of the sewerage system in the cities is extremely critical, and there are no sewerage systems in the rural areas. There is an urgent need for substantial improvements and extensions to the sewerage system treatment plants. The further development of water resources for agriculture and domestic use is one of the priorities to improve the agricultural yield of the country and the domestic and industrial demands for water.

Chapter 1

INTRODUCTION

Sudan is the largest country of the African continent, with an area of approximately one million square miles (2.5 x 10^6 km^2). Sudan is a federal republic located in eastern Africa. It extends between latitudes 3°'N and 23°'N, and longitudes 21° 45'E and 39°'E. Sudan is a relatively sparsely populated country. The total population, according to the 2006 census, was 39 x 10^6 inhabitants. The growth rate is 2.8%/y, and population density is 14 persons per square kilometer. The country is divided into 26 states and a federal district, in which the capital, Khartoum is located. Sudan is known as a country of plentiful water, rich in land, with highest total and renewable supply of fresh water in the region (eastern Africa). Sudan is considered one of the least developed countries, with a per capita income of less than US $400, and a real growth rate of 0.2 % of real gross domestic product (GDP) during the last ten years. However, during 1980s, the real growth rate of GDP was negative, mainly due to drought and desertification. The backbone of Sudan's economy is its agricultural sector. The agricultural sector determines to a great extent the economic performance of the Sudanese economy (Appendix 2). In fact, the country can be rescued by proper organization and utilization of its agricultural potential. Recent development due to rehabilitation and improvement in the agricultural sector has raised the share to 41%.

Agriculture continues to play a pivotal role in economy. It is directly influences the level of activities of all other sectors. It provides 90% of the raw materials for local industries, accounts for 80% of export earning, and provides income and employment for more than 80% of the population. The agricultural sector is composed of three distinct modes: Irrigated agriculture, mechanized rain-fed agriculture, and traditional rain-fed agriculture and livestock raising. Sudan is an agricultural country with fertile land, plenty of water resources, livestock, forestry resources, and agricultural residues. To ensure a better quality of life for all people, now and in the future, through the implementation of sustainable development initiatives that promote: (1) Food and water security (2) Economic efficient that helps to eliminate inequalities (3) Social equity for all, regardless of race, gender, disability or creed (4) Effective education for environmentally and socially responsible citizenship (5) Environment integrity, and environmental justice (6) Democracy, and mutual understanding between people.

In a country that is relatively sparsely populated, there are extreme pressures on water and waste systems, which can stunt the country's economic growth. However, Sudan has recognized the potential to alleviate some of these problems by promoting renewable water and utilizing its vast and diverse climate, landscape, and resources, and by coupling its solutions for waste disposal with its solutions for water production. Thus, Sudan may stand at the forefront of the global renewable water community and presents an example of how non-conventional water strategies may be implemented. In Sudan, more than ten million people do not have adequate access to water supply, twenty million inhabitants are without access to sanitation, and very little domestic sewage is being treated. The investment needed to fund the extension and improvement of these services is great. Most governments in developing countries are ready to admit that they lack the financial resources for proper water and sanitation schemes. Moreover, historically, bilateral and multilateral funding account for less than 10% of total investment needed. Thus, the need for private financing is imperative. Water utilities in developing countries need to work in earnest to improve the efficiency of operations. These improvements would not only lead to better services but also to enhanced net cash flows that can be re-invested to improve the quality of service. Staff productivity is another area where significant gains can be achieved. Investment and consumption subsidies have been predicated on the need to (a) help the poor, who do have not an access to basic services and (b) improve the environment. Failure of subsidies to reach intended objectives is, in part, from lack of transparency in their allocation. Subsidies are often indiscriminately assigned to support investment programs that benefit more middle- and high-income families that already receive acceptable service. Consumption subsidies often benefit upper-income domestic consumers much more than low-income ones. Many developing countries (Sudan is not an exception) are encouraging the participation of the private-sector as a means to improve productivity in the provision of water and of wastewater services.

Private-sector involvement is also needed to increase financial flows to expand the coverage and quality of services. A key element to successful private participation is the allocation of risks. How project risks are allocated and mitigated determines the financial and operational performance and success of the project, under the basic principle that the risk should be allocated to the party, which is best able to bear it. Many successful private-sector interventions have been undertaken. Private operators are not responsible for the financing of works; nonetheless, they can bring significant gains in productivity, which would allow the utility to allocate more resources to improve and extend services. Redressing productivity, subsidy and cross-subsidy issues before the private-sector is invited to participate has proven to be less contentious. I have previously sought to encourage more private-sector involvement (Omer, 1995).

Sudan is geo-politically well located, bridging the Arab world to Africa. Its large size and extension from south to north provides for several agro-ecological zones with a variety of climatic conditions, rainfall, soils and vegetation. These agro-ecological zones and the large mass of arable land support a variety of food and industrial crops. Vast natural pastures and forests support a large herd of livestock including cattle, sheep, goats and camels. It has plenty of solar radiation and vast productive land estimated at 84×10^6 hectares while the actual area under crops has not exceeded 12%. Water resources available to Sudan from the Nile system, together with groundwater resources, provide a potential for a threefold increase in the irrigated sub-sector (Omer, 1995). There are also opportunities for increased hydroelectrical generation. The strategy of Sudan at the national level aims at the multi-purpose use of water resources to ensure water security for attaining food security, drinking-water security, fiber-security, hydro-energy security, industrial security, navigation, waste disposal and the security at the regional levels within an environmentally sustainable development context and in harmony with the promotion of basin-wide integrated development of the shared water resources (Noureddine, 1997).

The government has continued to pay for the development and operation of water systems, but attempts are being sought to make the user communities pay water charges. In order to ensure the sustainability of water supplies, an adequate institutional and legal framework is needed. Funds must be generated (a) for production, (b) for environmental protection to ensure water quality, and (c) to ensure that water removal remains below the annual groundwater recharge. At present, there are private-sector providers who do not have an enabling environment to offer the services adequately. There is a need for the government to have a mechanism to assist in the regulation and harmonization of the private-sector providers. Privatization is part of a solution to improve services in delivery of water and of sanitation sector. At present, there is a transitional situation characterized by: (i) A resistance to payment of water charges; (ii) Insufficient suitable law and inadequate law enforcement; (iii) Insufficient capacities to provide services; and (iv) Inadequate interaction between the private citizen, business opportunities, and the government.

Sudan is known as a country of plentiful water, rich in land, with highest total and renewable supply of fresh water in the region (eastern Africa). The cultivable area is estimated at about 105 million ha (42% of the total land area), while in 2002 the cultivated land was 16.65 million ha (7% of the total land area and 16% of the cultivable area), comprising 16.23 million ha arable land and 0.42 million ha under permanent crops. The forest resources of Sudan cover approximately 27% of the total country's area. The main forest types include i) arid and semi-arid shrubs; ii) low rainfall savannah; iii) high rainfall savannah; iv) special areas of mountainous vegetation in Jebel Marra, the Red Sea Hills and the Imatong Mountains. Rangelands cover about 117 million ha. They spread over most ecological zones: the desert in the north, the semi-desert, the low rainfall savannah and the high rainfall woodlands in the south. Annual herbaceous plants with scattered trees and bushes dominate the northern rangelands. In the southern part, perennial herbaceous plants increase with dense stands of woody cover. The livestock population includes camels, sheep and goats, which are raised in the desert and semi-desert, and cattle that are raised in the medium rainfall savannah and in the flood plain of the Upper Nile. Almost all livestock is raised under nomadic and semi-nomadic systems. The country has a diverse and fairly rich wildlife. Of the 13 African mammalian orders, 12 are present in Sudan. The protected wildlife areas cover around 36 million ha. There are eight national parks, 13 game reserves and three sanctuaries.

This study comprises a comprehensive review of water sources, the environment and sustainable development. It includes the renewable water resources, water conservation scenarios and other mitigation measures necessary to reduce climate change. In fact, the country can be rescued by proper organization and utilization of its agricultural potential. Agriculture continues to play a pivotal role in economy.

Chapter 2

WATER RESOURCES

Water is one of the most fundamental of natural resources that a country must harness in its efforts for rapid economic development. The role of water in the development process cannot be over-emphasized. Sudan is rich in water (from the Nile system, rainfall and groundwater) and lands resources (Table 1). Surface water resources are estimated at 84 billion m^3. The annual rainfall varies from almost nil in the arid hot north to more than 1,600 mm in the tropical zone of the south. The total quantity of renewable groundwater is estimated to be 260 billion m^3, but only 1% of this amount is being utilized. Internal renewable water resources (IRWR) include the average annual flow of rivers and the recharge of groundwater (aquifers) generated from endogenous precipitation-precipitation occurring within a country's borders. IRWR are measured in cubic kilometers per year (km^3/year). Since data were collected in different years, they may not be directly comparable.

Surface water produced internally includes the average annual flow of rivers generated from endogenous precipitation and base flow generated by aquifers. Measuring or assessing total river flow occurring in a country on a yearly basis usually computes surface water resources (Table 2).

Groundwater recharge is the total volume of water entering aquifers within a country's borders from endogenous precipitation and surface water flow. Groundwater resources are estimated by measuring rainfall in arid areas where rainfall is assumed to infiltrate into aquifers. Where data are available, groundwater resources in humid areas have been considered as equivalent to the base flow of rivers.

Overlap is the volume of water resources common to both surface and groundwater. It is subtracted when calculating IRWR to avoid double counting. Two types of exchanges create overlap: contribution of aquifers to surface flow, and recharge of aquifers by surface run-off. In humid temperate or tropical regions, the entire volume of groundwater recharge typically contributes to surface water flow. In karstic domains (regions with porous limestone rock formations), a portion of groundwater resources is assumed to contribute to surface water flow. In arid and semi-arid countries, surface water flows recharge groundwater by infiltrating through the soil during floods. This recharge is either directly measured or inferred by characteristics of the aquifers and piezometric levels.

Table 1. Land use, land-resource zones and water resources

(a) Land use (x 10^6 hectares)

Geographical area	250.6
Land area	237.6
Cultivable area	8.4
Pastures	29.9
Forests and woodland	108.3
Uncultivatable land	81.0
Area under crop	10.0

(b) Land-resource zones

Zone	Area as % to total area of Sudan	Persons per km^2	Mean average rainfall range (mm)
Desert	44	2	0-200
QOS sands (sand dunes)	10	11	200-800
Central clay plains	14	19	200-800
Southern clay plains	12	8	800-900
Ironstone plateau	12	7	800-1400
Hill area and others	8	16	Variable

(c) Water resources

Water resource	Available number	Static water level (m)	Number
Haffirs* (excavated land)	824	0-0	824
Slow sand filters			
Open shallow wells	128	0-0	128
Boreholes (deep wells)	3000	0-10	3000
	2259	0-25	1248
		26-50	478
		51-75	287
		76-100	246

* Haffirs are excavated land to store rainwater for use in drought seasons (Appendix 4).

(d) Geological formations

Basins	Amount of water recharged (10^6 m^3)	Water level below land (m)	Aquifer thickness (m)	Velocity (m/year)	Abstraction (10^6 m^3/year)
Sahara Nile	136	30-100	300-500	1-2.5	7.3
Sahara Nubian	20.6	10-50	300-500	0.8-1.5	1.5
Central Darfur	47.6	25-100	250-550	0.3-6.0	5.5
Nuhui	15.4	75-120	200-400	1.0-2.75	1.6
Sag El Na'am	13.5	50-1000	300-500	1.0-25.0	2.5
River Atbara	150	100-150	250-300	0.3-5.0	2.3
Sudd	341	10-25	200-400	0.1-1.8	1.8
Western Kordofan	15	50-70	300-500	0.1-0.3	1.7
Baggara	155	10-75	300-500	0.1-2.4	11.9
Blue Nile	70.9	10-50	250-500	0.1-2.5	10.2
The Alluvial	N.A	Shallow	N.A	N.A	N.A
Gedaref	41.7	50-75	200-500	0.1-2.0	1.2
Shagara	1.1	25-30	200-300	0.1-2.5	0.7

(e) Annual rainfall pattern (10^9 m^3 $year^{-1}$)

Average range (mm)	Annual rainfall	Percent (%)
< 100	41.7	3.8
100-300	76.5	6.9
300-600	199.5	18.2
600-1000	515.5	47.2
> 1000	261.0	23.9
Total	1094.2	100.0

Total internal renewable water resources is the sum of surface and groundwater resources minus overlap; in other words, IRWR = Surface water resources + Groundwater recharge – Overlap. Natural incoming flow originating outside a country's borders is not included in the total.

Per capita internal renewable water resources (IRWR) are measured in cubic meters per person per year (m^3/person/year). Per capita values were calculated by using national population data for 2001.

Natural renewable water resources, also known as Actual Renewable Water Resources, are the sum of internal renewable water resources and natural flow originating outside of the country. Natural renewable water resources (NRWR) are computed by adding together both internal renewable water resources (IRWR) and natural flows (flow to and from other countries). Natural incoming flow is the average amount of water, which would flow into the country without human influence. In some arid and semi-arid countries, actual water resources are presented instead of natural renewable water resources.

Per capita natural renewable water resources are measured in cubic meters per person per year (m^3/person/year). Per capita values were calculated by using national population data for 2002.

Internally produced water resources in Sudan are rather limited. The erratic nature of the rainfall and its concentration in a short season places Sudan in a vulnerable situation, especially in rain-fed areas. Surface water in Sudan comprises the Nile river system (Nilotic water) and other, Non-Nilotic streams. Sixty-four percent of the Nile Basin lies within Sudan, while 80% of Sudan lies in the Nile Basin. Local rainfall is the main source of the Non-Nilotic streams and of the Bahr el Ghazal basin, whereas rainfall over the Central African Plateau (Equatorial Lakes) and over the Ethiopian-Eritrean highlands is the main source of the Nile River system and other transboundary seasonal streams (Gash and Baraka). Sudan shares parts of the following basins with neighboring countries:

- The Nile Basin, 1,978,506 km^2 (79% of the area of the country);
- The Northern Interior Basins, covering 313,365 km^2 in the northwest part of the country (12.5%);
- The Lake Chad Basin, in the west of the country, along the border with Chad and the Central African Republic, covering 101,048 km^2 (4%);
- The Northeast Coast Basins, representing a strip along the Red Sea coast of the country, covering 96,450 km^2 (3.8%);
- The Rift Valley Basin, in the southeast part of the country at the border with Ethiopia and Kenya, covering 16,441 km^2 (0.7%).

The Nile system within Sudan comprises:

- The Blue Nile, Sobat and Atbara Rivers originating in the Ethiopian highlands;
- The White Nile system, upstream of Sobat River, originating on the Lakes Plateau;
- The Bahr el Ghazal Basin, an internal basin in southwest Sudan.

Table 2. Water resources and freshwater ecosystems

Internal renewable water resources (IRWR), 1977-2001, in km^3	Sudan	Sub-Saharan Africa
Surface water produced internally	28	3812
Groundwater recharge	7	1549
Overlap (shared by groundwater and surface water)	5	1468
Total IRWR (surface water + groundwater – overlap)	30	3901
Per capita IRWR, 2001 m^3	921	5705
Natural renewable water resources include flows from other countries	65	-
Total 1977-2001 km^3	1981	-
Per capita 2002 m^3/person		
Annual river flows:	119	-
From other countries km^3	-	-
To other countries km^3		
Water withdrawals		
Year of withdrawal data	1995	-
Total withdrawals km^3	17.8	-
Withdrawals per capita m^3	637	-
Withdrawals as a percentage of actual renewable water resources	32.1%	-
Withdrawals by sector (as a percent of total)		
Agriculture	94%	-
Industry	1%	-
Domestic	4%	-
Desalination (various years)		
Desalinated water production (million m^3)	0	-

The characteristics of the Nile system tributaries are the following:

- *The Blue Nile*: The flow of the Blue Nile reflects the seasonality of rainfall over the Ethiopian highlands where the two flow periods are distinct. The flood period, or wet season, extends from July to October, with the maximum in August to September, and low flow or dry season from November to June. Therefore, the annual Blue Nile hydrograph has a constant bell-shaped pattern, regardless of variation in the annual flow volumes. The average annual flow of the Blue Nile and its tributaries upstream of the confluence with the White Nile at Khartoum is about 50 km^3; the daily flow fluctuates between 10 million m^3 in April to 500 million m^3 in August (ratio of 1:50).
- *The White Nile*: Due to losses in the Sudd swamp area, the White Nile leaves this area with only about 16 km^3, out of 37 km^3 on entering it. The river receives about 13 km^3 from the Sobat River before joining the Blue Nile at Khartoum. The contribution of the Bahr el Ghazal

basin is negligible, estimated at about 0.5 km^3. The average annual flow of the White Nile System at Malakal is about 29.5 km^3, and the daily discharge fluctuates between 50 million m^3 in April, to 110 million m^3 in November (ratio 1:2). During the flood period, the Blue Nile forms a natural dam that obstructs the flow of the White Nile and consequently floods the area upstream of the confluence.

- *The Atbara River*: This is a highly seasonal river, with an annual flow upstream of its confluence with the Nile of about 10 km^3 restricted to the flood period of July to October, the maximum occurring between August to September. The river has a steep slope and small catchments and reflects the rainfall over the upper catchments as runoff at Sudan border within one to two days.
- *The Main Nile*: The reach of the Nile downstream of the confluence of the Blue Nile and the White Nile Rivers is known as the Main Nile. The Atbara River is regarded as the only and last tributary joining the Main Nile. The average annual flow of the Main Nile at the Sudan-Egypt border at Aswan is estimated at 84 km^3.

The average annual yield of the Non-Nilotic streams is estimated at about 7 km^3/yr, of which 5 km^3/yr are internally produced. The major streams are the Gash and Baraka in the east of the country, both of which are characterized by large variations in annual flow and heavy silt loads.

The major groundwater formations and basins are the Nubian Sandstone Basin and the Umm Ruwaba Basins. The Ghazal, Sudd and Sobat swamps in the south of the country represent major wetlands, from which evaporation is exceptionally high. According to an estimate from 1980, the extent of the Sudd is over 16,200 km^2, but the surface area fluctuates with rainfall. Sudan's total natural renewable water resources are estimated to be 149 km^3/yr, of which 30 km^3/yr are internally produced (Table 3). In a tenth frequency dry year, the internal water resources are reduced to about 22.3 km^3/yr. Of the internal water resources, 28 km^3/yr are surface water and 7 km^3/yr are groundwater, while the overlap between surface water and groundwater is estimated at 5 km^3/yr. As a result of the Nile Waters Agreement with Egypt, total actual renewable water resources of the country amount to 64.5 km^3/yr.

The high variability of river flows necessitates storage facilities. The total storage capacity of the following four main dams is estimated at 8.73 km^3, reduced to about 6.90 km^3 owing to sedimentation and debris:

- The Sennar Dam on the Blue Nile (design capacity 0.93 km^3, present capacity 0.60 km^3) is for the flood control and irrigation of the Gezira Scheme.
- The Roseires Dam on the Blue Nile (design capacity 3.0 km^3, present capacity 2.2 km^3; there are plans to increase the present dam height of 60 m to provide an extra capacity of 4.0 km^3) is for flood control and utilizes part of the country's share of the Nile waters for irrigation.
- The Jebel Aulia Dam on the White Nile (design capacity 3.5 km^3, present capacity 3.5 km^3) was originally designed to benefit Egypt by augmenting the supply of summer flow to the Aswan dam. After the construction of the High Aswan Dam, it was no longer needed by Egypt and was officially handed over to the Sudan in 1977.
- The El Girba Dam on the Atbara River (design capacity 1.3 km^3, present capacity 0.6 km^3) is for flood control, irrigation of New Halfa Scheme for the benefit of the people displaced by the High Aswan Dam, and hydropower.

A small barrage was constructed on the Rahad River to divert floodwater to the Rahad Agricultural Scheme and to siphon underneath the Dinder River to augment the water supply during the dry season from the Meina Pump Station on the Blue Nile. The Jonglei Canal, between Bahr el Jebel and the White Nile, was planned to divert water from upstream of the Sudd to a point farther down the White Nile, bypassing the swamps, to make more water available for use downstream. Works on it were discontinued in 1983, after two-thirds were completed. Non-conventional water sources are limited in Sudan. However, the desalination of seawater was introduced recently in Port Sudan town. Fossil groundwater resources are estimated to be 16,000 km^3.

Total water withdrawal in the Sudan was estimated at 37 km^3 for the year 2000 (Table 3). The largest water user by far was agriculture with 36 km^3 (Figure 1). The domestic sector and industry accounted for withdrawals of 0.99 km^3 and 0.26 km^3, respectively. Water used in Sudan derives almost exclusively from surface water resources. Groundwater is used only in very limited areas and mainly for domestic water supply.

Table 3. Water sources and uses

Renewable water resources			
Average precipitation		416	mm/yr
		1042	$10^9m^3/yr$
Internal renewable water resources		30.0	$10^9m^3/yr$
Total actual renewable water resources		64.5	$10^9m^3/yr$
Dependency ratio		76.9	%
Total actual renewable water resources per inhabitant	2004	1879	m^3/yr
Total dam capacity	1995	8730	10^6m^3
Water withdrawal			
Total Water withdrawal	2000	37314	10^6m^3
-Irrigation+Livestock	2000	36069	10^6m^3
-Domestic	2000	987	10^6m^3
-Industry	2000	258	10^6m^3
- per inhabitant	2000	1187	m^3/yr
- as % of total actual renewable water resources	2000	58	%
Non-conventional sources of water			
Produced wastewater		-	10^6m^3
Treated wastewater		-	10^6m^3
Reused treated wastewater		-	10^6m^3
Desalinated water produced	1990	0.4	10^6m^3
Reused Agricultural drainage water		-	10^6m^3

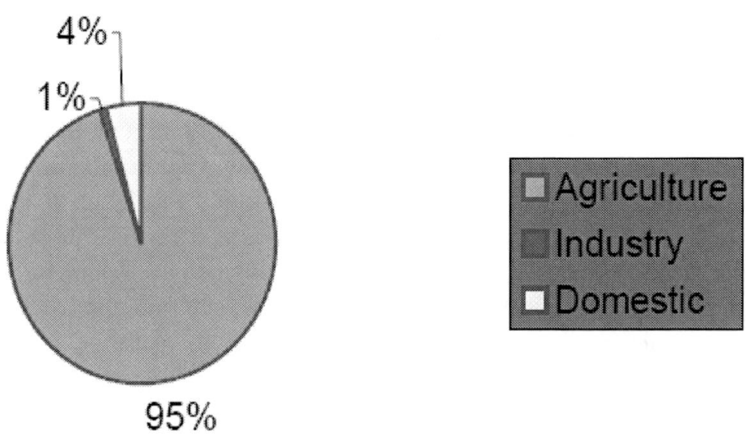

Figure 1. Surface water withdrawals by sector in Sudan (total 37.314 km^3 in 2000).

Water-resources assessment in Sudan is not an easy task because of the uncertainty of parameters, numerous factors that affect the variables, lack of information and inaccurate measurements. However, according to seasonal water availability, Sudan can be divided into three zones: (a) areas with water availability throughout the year are the rainy regions (equatorial tropical zones); (b) areas with seasonal water availability; and (c) areas with water deficit throughout the year. The last occupy more than half the area of Sudan (Figure 2). Water resources in Sudan include surface water and ground water spread over large parts of the country. Both resources are utilized for agricultural production by the use of canal system from dams, pumps, embankments (flood), and wells.

2.1. The Nile System

The Nile system travels Sudan from south to north. The Nile basin system in Sudan comprises the Blue Nile, Rahad, and Dender system. The White Nile, Bahr El Gabal, Zaraf, Bahr El Ghazal and Sobat, and in addition, the Atbara system.

Bahr El Gabal	26×10^9 m^3
Bahr El Ghazal	15×10^9 m^3
Sobat at Malakal	13×10^9 m^3
Total at Malakal	54×10^9 m^3
Losses in swamps	27×10^9 m^3
White Nile at Malakal	26×10^9 m^3
Blue Nile at junction	53.4×10^9 m^3
Atbara system	11.6×10^9 m^3
Losses along river	77×10^9 m^3
Net available at Aswan	84×10^9 m^3

In 1959, the Nile water agreement was concluded between Egypt and Sudan. The agreement is based on average Nile yield of 84×10^9 m^3 (measured at Aswan). After construction of the High Aswan Dam (saving 22×10^9 m^3): 14.5×10^9 m^3 for Sudan and 7.5×10^9 m^3 for Egypt. As a result, Sudan share has become 18.5×10^9 m^3 and 55.5×10^9 m^3 for Egypt, leaving 10^{10} m^3 for evaporation from the Aswan High Dam reservoir.

2.2. Ground Water

The ground water quality is suitable for animal and human consumption as well as for agriculture and other uses. The potential renewable ground water suitable is estimated at 6 x 10^9 m^3. The main aquifer is Nubian sandstone, covering 28% of surface of the country. Groundwater is mainly used to satisfy animal and human needs in rural areas, while small areas are irrigated from groundwater. About 5 x 10^9 m^3 may be used for agriculture during the coming ten years. Current use for both agriculture and drinking water is about 1.2 x 10^9 m^3. The groundwater potential is believed much more than presently estimated.

The country has an agricultural potential of 105 million ha, of which only 16.7 million ha are cultivated and only about 1.9 million ha out of an irrigation potential of 2.8 million ha are under irrigation now. Therefore, there is ample room for further developments, especially in the irrigation sub-sector. However, there are three major constraints to irrigation development in Sudan:

- The ineffective process of annual maintenance of civil works (reportedly due to lack of funds), especially for the removal of silt and weeds from irrigation canals, which slows down water flow and causes continuous shrinkage of the actual irrigated area;
- The steady increase in development costs, which has been aggravated by the continuous devaluation of the local currency;
- The lack of farmer involvement in the planning and operation of the schemes and related services.

Generally, the water supply for all of the irrigation schemes is provided by dams and/or pumps and extensive networks of canals covering the whole schemes as well as drainage networks of canals. The overall objective of water management policies is to improve water use efficiency in agriculture, which includes efficient control of water in the irrigation networks, maintenance of the irrigation structures, provision of technical capacities capable to operate the systems, and efficient and economical maintenance of the irrigation systems. Supplementary irrigation could increase the very low or zero productivity of crops and fodder. The conjunctive use of groundwater and surface water could help to optimize the water resource productivity.

Chapter 3

USES OF WATER

The utilization of water in Sudan is widely estimated in agriculture, human use, domestic, animal uses, industrial, hydropower generation and navigation. The agricultural sector is the major source of water consumption in Sudan. Sudan is presently utilizing 16.5 x 10^9 m^3 annually from its share in irrigated agriculture sub-sector, currently covering an area of 1.7 x 10^6 hectares. The potential is threefold. The irrigated crops include cotton, wheat, sorghum, groundnuts, sugarcane, vegetables and fruits. Mechanized and traditional rain-fed farming sectors cover an area estimated annually at about 8-10 x 10^6 hectares. The total area cropped in the rain-fed areas varies from year to year, depending on rainfall. Crops grown in the rainfall sub-sector include sorghum, millet, sesame, sunflower, and groundnuts. The potential in the rainfall sector is more than fivefold. The southern tropical zone of Sudan is distinct with red lateritic soil and heavy rainfall ranging between 1,000 to 1,500 mm over seven to eight months. Agriculture is largely subsistence with a wide range of food crops including maize, sorghum, millet, root crops, banana, pulses, tea, and coffee. Because of the marked fluctuation between the flood discharge and the low season period in the Nile system, storage reservoirs in Sennar, Rosaries, Griba and Jebel Aulia were constructed to ensure the availability for water during the recession period. These dams are used for irrigation, navigation and hydropower generation. Drinking water supplies have been provided for people and animal in most of the urban and rural areas, but still more than 40% of the rural areas and more than 25% in urban areas are in need of safe drinking water supplies. Water use for industry and sanitation is still very limited. The demand for water in industries at the present is estimated at about 0.24 x 10^9 m^3 per year, most of which is from the surface water. The most important research and development policies that have been adopted in different fields of water resources are (i) the water resource; (ii) irrigation development; (iii) the re-use of drainage water and groundwater; (iv) preventive maintenance of existing facilities, including canals; (v) aquatic weed control and river channel development, and (vi) protection plans for water resources in general. The base of physical and human resources can provide for sustainable agriculture growth and food security for itself and for others in the region. Failure to plan for sustainability in the past derives from several causes and constraints, which are manageable. These include misguided policies, poor infrastructure, low level of technology use, recurring droughts and political instability. Perhaps the biggest challenge is that of finding resources for capital improvements in the light of changing water-quality regulations and ageing systems (James, 1994). Environmental institutions and regulators are already working to solve the next set of issues in their constant pursuit of better quality of live. Increasing watershed and source protection that combat microbial and organic contaminations is possible with new detection techniques that enable identification and control of pesticide runoff, and reduce chlorine by-products. Upgrades of ageing infrastructure are receiving unprecedented attention. Surface-water systems are continuing to rely on conventional treatment (coagulation, flocculation, sedimentation, filtration and disinfection) for particulate removal, but other treatment processes such as ozone and granular activated carbon will see increased use, while new disinfection strategies will be needed to minimize both microbial risks and unwanted by-products. Sudan is, therefore, moving into a new era in the protection of drinking-water supplies. It is now time for water utilities to combine creative management, dependable treatment methods and new technologies to ensure that drinking water is as safe as possible. The demand for water in Sudan has increased tremendously over the years and will continue to increase in view of the accelerating pace of population growth, urbanization and industrialization. There are many approaches and programs through which human resource can be properly developed for the water sector. These programs are not mutually exclusive and include (1) Non-formal education through media and non-governmental groups. (2) Formal education through universities and technical colleges. (3) On-the-job training and regular refresher courses.

Due to excess rainfall and sometimes to misuse of irrigation water, all the irrigation schemes need drainage networks to remove any excess water from the cultivated areas. In low areas, minor drains and collector drains are constructed to remove this excess water by gravity into cutout low areas or into natural drains. Sometimes, pumps are used to take water from low lands into areas outside the scheme. Also, escape drains are constructed along the main canal to carry any excess water to the nearest river or natural drain. It is estimated that in 2000, about 560,000 ha were drained. The development plans for the irrigation sector include the rehabilitation of the existing irrigation schemes, a shift of emphasis towards the development of small-scale irrigation schemes, and phased development and vertical expansion.

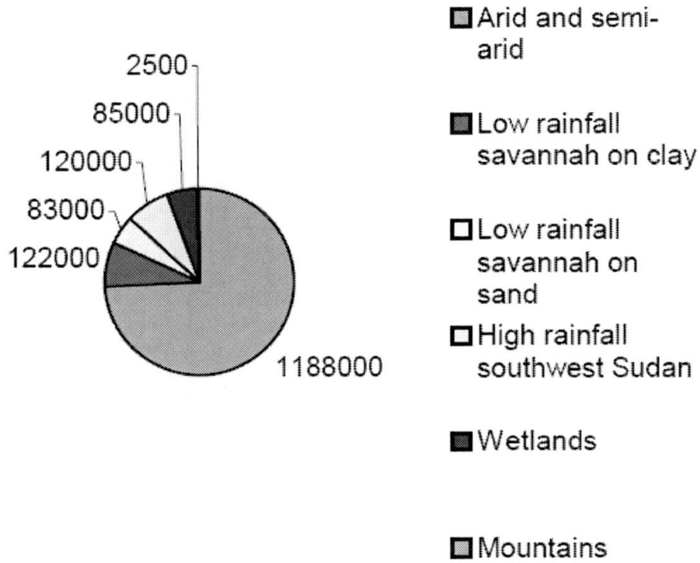

Figure 2. shows the major Sudan's ecosystems (km^2).

The main irrigated crops are sorghum, cotton, fodder, wheat, groundnuts and vegetables. Other crops under irrigation are sugar cane, maize, sunflower, potatoes, roots and tubers and rice. Irrigated agriculture has been Sudan's largest economic investment, yet returns have been far below potential. A study by the World Bank showed that, during the period of 1976 to 1989, yields were low and extremely variable, and cultivated areas suffered a gradual decline. A study undertaken in the Rahad Scheme based on data from 1977 to 1995, shows that actual crop yields are well below potential yields. The same study also estimated the water use efficiency and found an overall efficiency of 63 to 68%. The distribution efficiency of the network was 93%, and estimated field losses were 25 to 30%.

Chapter 4

WATER QUALITY AND SANITATION MANAGEMENT

Community water supply and sanitation management is a new form of cooperation between support agencies in the water and sanitation sector and communities. It involves a common search to identify problems with the local water supply and sanitation systems and the possibilities for, and constraints on, management by communities, and to find possible solutions that may be tested. Some fundamental principles of community water and sanitation management are (i) Increased management capacities as the basis for improved water and sanitation systems, wherein each community develops its own specific management systems; and (ii) Communities' ownership of the process of water charges; facilitators and local researchers participate in the community's projects, not the other way around. Through that approach, the support agency is no longer the provider of technical goods or solutions, but the facilitator of a process to enhance the capacity of the community to manage its own water and sanitation systems. Constraints include (i) A lack of funds or long delays in allocating funds for requirements such as operation and maintenance of irrigation and drainage projects; (ii) Deterioration of data-collection; (iii) A lack of appropriate and consistent policies for water development for both large- and small-scale projects; (iv) Serious delays in completing water projects after major investments have been made, e.g., dams, other hydraulic structures, and main canals are finished; (v) An absence or inadequacy of monitoring, evaluation, and feedback at local, national and international levels; (vi) A lack of proper policies on cost recovery and for water pricing or, if policies exist, lack of implementation; (vii) A shortage of professional and technical personal, and training facilities; (viii) A lack of beneficiary participation in planning, implementation, and operation of projects; (ix) Inadequacy of knowledge, and absence of appropriate research to develop new technologies and approaches, and an absence of incentives to adopt them; (x) General institutional weaknesses and lack of coordination between irrigation, agriculture, energy, human health services, environment, and planning; (xi) Inappropriate project development by donor agencies, e.g., irrigation development with drainage, supporting projects that should not have been supported; and (xii) A lack of donor coordination resulting in differing approaches and methodologies and thus conflicting advice.

As developing nations strive to provide a safe and reliable drinking-water supply to their growing and increasingly urbanized populations, it is becoming more evident that new approaches to this problem are needed. To meet this challenge, new methods of reclaiming and re-using water have been developed in cost-effective and environmentally sound ways (ODA, 1987, Seckler, 1992;, Salih et al., 1992). Despite the constraints, over the last decade, the rate of implementation of rural and peri-urban water supply and sanitation programs has increased greatly, and many people are now being served more adequately. The following are Sudan's experiences in water supply and sanitation projects:

At the community level:

- Participatory approaches in planning, implementation and monitoring.
- Establishment and training of water committees.
- Clear ownership of improved water supply and sanitation systems.
- Selection of technology and level of service by consumers.
- Sensitive timing of education in hygiene and sanitation.
- Establishment and training of reliable financial and maintenance management.

At the district and national level:

- Integrated and multi-sectoral approach to development.
- Training and material development for district and extension staff.
- Continuing support from well-trained, integrated and multi-sectoral extension teams.
- Establishment of technical support system.
- Multi-sectoral advisory group to include representation from membership training and research institutions.

- Development and dissemination of relevant information to district and extension staff.

Water is a substance of paramount ecological, economical, and social importance. Interrelationships inherent in water use should encourage integrated management. To better manage water resources, the following are needed: (1) Ensure more reliable availability of water and efficient use of water in the agricultural sector. (2) Mitigate flood damage by ground reservoirs and terraces. (3) Control water pollution. (4) Prevent development of soil salinity and water logging. (5) Reduce the spread of water-borne diseases.

The development of new, modern, and complete water resources information systems is one of the basic needs for the implementation of the water resources management system in Sudan. Decisions relating to drought, flood or over-exploitation can only be correct if based on reliable information. A comprehensive database on water availability, users, water quality monitoring, current technologies (like geographical information systems) is certainly the way to produce an efficient framework for decision-making. Lack of information is one of the most critical points regarding global development and implementation of new management systems (FAO, 1999). For example, the types of data related to flood management in Sudan and globally include (i) Topographic data (elevations, land use, and soils). (ii) Imagery (satellite images). (iii) Administrative data (political boundaries). (iv) Infrastructure data (roads, levels, wells, utilities, bridges and culverts, hydraulic structure). (v) Environmental data (threatened and endangered species, critical aquatic and wildlife habitat). (vi) Hydrometeorology data (flood plain delineation, stream flows, precipitation, temperature, wind, solar radiation, soil moisture, humidity, discharge rating curves, and flood frequency). (vii) Economic data (stage-damage relationships), and (viii) Emergency management data (emergency plans, and census data).

The emerging water crisis, in terms of both water quantity and quality, requires new approaches and actions. Priority areas needing concerted action in various sectors in Sudan are (a) Water use efficiency, (b) Flood control, (c) Water resources management, (d) Water quality management and provision of safe drinking water, and (e) Coordination and integration of various aspects of water management, and water management with other related resources and societal concern. The following are recommended:

- The community must be the focus of benefits accruing from restructures. For the most legal framework to protect community interest on the basis of equity and distribution, handover of the assets to the community should be examined. Communities should encourage the transfer of the management of water schemes to a professional entity.
- The private-sector should be used to mobilize, and strengthen the technical and financial resources, from within and without the country to implement the services, with particular emphasis on utilization of local resources.
- The government should provide the necessary financial resources to guide the process of community management of water supplies. The government should be a facilitator through setting up standards, specifications and rules to help harmonize the private-sector and establish a legal, independent, body by an act of parliament to monitor and control the providers. The government should assist poor communities that cannot afford service cost and alleviate social-economic negative aspects of privatization.
- All sector actors should create awareness in the community of the roles of the private business and government in the provision of water and sanitation services.
- Support agencies should assist with financial and technical assistance, the establishment of training facilities, coordination, development and dissemination of water projects, and then evaluate of projects with a view to future improvements.

The institutional framework provides the basis by which all actions are taken, and an assessment of its functional character helps determine the collaborative potential. The resulting criteria for measuring a given community's institutional capacity can be found in Table (4). Sudan has to build a bigger and better base of science and technology, besides strengthening the national science and technology systems. Education and training play key roles in human resource development in water and other sectors.

4.1. Groundwater and Environment

In northern Sudan, the desert environment is fragile and highly affected by man's activities. One of the geotechnical hazards that may occur under desert conditions and of special interest in Sudan is sand-drifting and dune movement. Moving sand covers approximately one-third of the country. Because sand poses natural erosional-depositional hazards on the existing structures, such as roads and urbanized areas, it becomes necessary to study the behavior of the sand forms in the different parts of the country. Deserts are known to be simply barren areas; they are scientifically defined in terms of aridity, soil type, topography and vegetation. Most of the Middle Eastern countries lie within the semi-arid, arid, and hyper-arid desert zones, with an aridity index (ratio between annual precipitation and mean annual potential evapotranspiration) ranging between 0.03 and 0.02 (Anon, 1979). The heavy mineral content ranges between 11.8 and 15.4% (Brown, 1999). The desert environment, being a fragile ecosystem, needs to be treated with care. Intercommunications among different national and international agencies and education of the layman should help to keep the system balanced and reduce environmental hazards. In addition, any suggested remedial measures should be planned with nature in mind and be engineered with natural materials.

As a result of the different sources of pollution in the Khartoum State, together with the development of industries and private agricultural projects, a serious impact to the environment has been encountered. The landscape configuration changes, land subsidence, soil pollution and degradation, air pollution and surface and groundwater contamination are among the various environmental problems in the State. Toxic materials percolating downwards from wastewater disposal wells, oxidation ponds, septic tanks and chemical plants mainly cause groundwater aquifer pollution. The increasing population in the Khartoum State led to increasing demands for safe domestic water supply. Therefore, aquifer vulnerability assessment becomes prerequisite in order to identify aquifers with high pollution potential. The assessment can help to provide tools for groundwater protection and management.

Facing the increasing pressure on water resources in all regions of Sudan requires an urgent need for developing new tools to achieve their sustainable management by applying the principles of eco-hydrology. Generally, the eastern, southwestern banks of the White Nile and part of the northern bank of the Blue Nile are considered as more vulnerable to contamination and, consequently, need to be managed more closely. In fact, it is highly influenced by the characteristics of the local hydrological setting and the depth to water and slope parameters.

Table 4. Capacity assessment for community management: Institutional factors (Salih, 1992)

High capacity (plans, etc., in places)	• Basin-wide management plan has been drafted. • Natural mitigation strategy in place. • Basin-wide coordination and communications strategy instituted. • Trained emergency management staff coordinating at the regional level. • Effective regulatory policies that address floodplain occupancy. • Decentralised decision-making with a high degree of local autonomy.
Medium capacity (evidence of activity ongoing)	• Evidence of an updated national response plan. • Bilateral response agreements. • Evidence of regional preparedness and response training. • Some trained emergency management staff at the local and/or national level. • Evidence of some regulatory policies designed to address floodplain occupancy. • Attempts to decentralise decision-making, moderate local discretion.
Low capacity (no formalization in place nor apparently evolving)	• No existing flood response plan. • No evidence of mitigation-related activities. • Poor local-and national-level coordination and communications. • Little or no evidence of flood preparedness and response training. • No regulatory policies addressing floodplain occupancy. • Centralized decision-making, no evidence of local autonomy.

Chapter 5

WATER MANAGEMENT, POLICIES AND LEGISLATION RELATED TO WATER USE

5.1. INSTITUTIONS

The Ministry of Agriculture and Natural Resources (MANR) supervises the Agricultural Corporations that manage the large irrigation schemes, while the Ministry of Irrigation and Water Resources (MIWR) is responsible for delivering irrigation water. The Ministry of Irrigation and Water Resources (MIWR) is the federal body in Sudan legally responsible for all water affairs. It offers technical advice and assistance to water projects within the states and the private-sector. It is in charge of the groundwater, the Non-Nilotic streams and valleys under the Groundwater and Wadis Directorate. It undertakes its task in coordination with the relevant sectors, departments and technical offices (agriculture, industry, foreign, electricity, and investment, etc.). It has the following responsibilities:

- Satisfaction of the water requirements of the various users through the country;
- Water resources planning, management and development;
- International and regional cooperation concerning the shared water sources;
- Planning, design, execution, operation and maintenance of the different irrigation schemes;
- Control of water abstraction;
- Construction of new irrigation works;

- Operation and maintenance of all large-scale irrigation structures and drinking water facilities;
- Provision of the means for hydropower generation and protection of the water-related environment.

5.2. WATER MANAGEMENT

The semi-autonomous Sudan Gezira Board (SGB) manages the Gezira Scheme on a vertically integrated basis. The MIWR is responsible for managing the Sennar Dam on the Blue Nile and the upper reaches of the irrigation system, responding to requests for water delivery from SGB's field staff. Within the scheme, the SGB serves as landlord, operates and maintains the lower reaches of the irrigation system and provides most of the inputs and services required by farmers to produce cotton, which is transported by the Board to its ginneries and sold on behalf of growers by the Sudan Cotton Company Limited. The SGB recovers the cost of advances made for inputs and services from the cotton sales before payment is made to the farmer. Tenants are wholly responsible for growing other crops in prescribed rotations with cotton (sorghum, groundnuts, forage, wheat, and vegetables), making their own arrangements for input supplies and marketing. By 2001, in the Gezira Scheme, Minor Canal Committees had been formed along the minor irrigation canals, and representatives of each of these committees constitute the Irrigation Committee at the block level. In addition to the Irrigation Committee, a Financial Committee has been established that is coordinating the reimbursement of the seasonal credits and arrangements for procurement of new inputs. The Irrigation Committee, with representatives of each of the minor canal committees, will be responsible for the operation and maintenance of the minor irrigation system, a task presently entrusted to the SGB, with the Ministry of Irrigation responsible of supplying the main system.

To address some of the problems facing irrigation management and development, the government has formalized a policy framework that includes:

- Transferring the operation and production of large- and medium-size irrigation schemes to the farmers and giving them full responsibility for water management on the irrigation system below the minor canals level through establishment of voluntary water users associations (WUAs).

- Fostering sustainable productivity of the large schemes through rehabilitation, combined with financial and institutional reform.
- Grouping, rehabilitating and handing over the relatively small-size pump schemes in the Blue Nile and the White Nile. These schemes were originally established and run by the government. Recently, and in accordance with the economic reforms, these schemes were handed over to the private-sector represented by individual farmers, cooperatives or private companies.

5.3. Legislation

In 1992, the national economy was reoriented towards a free economy, a policy shift that impacted the agricultural sector profoundly. The government withdrew from the direct financing of agriculture, provision of inputs and services. The government, within its policy of withdrawal from provision of goods and services, handed over all the small- and medium-size irrigation schemes under its control to the farmers. The handing-over policy was not successful because farmers were ill prepared, and most of the schemes were in need of rehabilitation. Since 1992, the cropped areas and the productivity of many schemes have sharply declined.

Agricultural sectoral policies for the irrigated sub-sector include the following:

- To extend the market economy to all crops to allow the farmer to choose a suitable crop mix.
- To provide support to the agricultural research institutions so as to explore suitable technologies for improving crop production and productivity. In addition, to provide support to agricultural extension and services.
- To encourage the private-sector to provide agricultural inputs for the agricultural sector.
- To encourage exports through improvement of quality so as to meet international standards.
- Establishment of specialized crop committees for main crops like cotton, with the objective of achieving all necessary coordination between the concerned authorities.
- Farmers participate in agricultural policy formulation.

The Sudan Comprehensive National Strategy for the Agricultural Sector (1992 to 2002) put food security, sustained agricultural development, efficient resource utilization and yield enhancement on the top of the agenda. Little has been done, however, to improve the accessibility to food of the poor, the vulnerable and the marginalized strata of society.

5.4. Finances

Financing irrigation operation and maintenance (O&M) through fees collected from the beneficiaries of the irrigation system was first introduced in Sudan with the introduction of the modern irrigation system at the El Zeidab scheme in 1909, when a private foreign company erected a pump station to irrigate local farmers' land for an agreed irrigation fee. After the success of the experiment for the first two seasons, a bad crop yield in 1911/1912, meant that the farmers were unable to pay their irrigation fees. The company experienced heavy losses and decided to pull out of the scheme.

Experience from El Zeidab scheme was used in selecting the form of production relationship among the government, the Sudan Plantation Syndicate and the farmers when the Gezira scheme was commissioned. To avoid the inability of some farmers to pay irrigation fees in case of bad crop yields, a "sharing system" between the three parties was adopted. This system continued until 1981, when it was replaced by what is known as the "individual account system," in which each individual farmer is treated separately in terms of cost and profit. The objective was to create some incentive for the individual farmers to increase their productivity. The new account system failed to achieve break-even productivity. The individual account system was also applied in all the irrigation schemes run by the government at that time. Payment of irrigation fees by the farmers continued in all government schemes from 1981 to 1995. During this period, irrigation fees collected were very low, averaging about 50% only. The non-recovered part of the water supply costs is borne by the government.

Starting from 1995, and as part of the liberalization of the economy, the government withdrew from financing the cost of irrigation services, among other things. Farmers were left to pay irrigation fees to the newly established Irrigation Water Corporation (IWC), which uses these fees directly to provide water supply services to the farmers. Instead of the IWC setting up its own mechanism for collecting the fees directly from the farmers, it relies on the Agricultural Corporations (ACs) managing the scheme to collect the fees from the farmers. Because these ACs were also facing considerable financial difficulties, part of the water fees collected may not reach the IWC, and part of the collected fees paid to IWC is delayed for sometime as it is used for financing other urgent activities. The result of this is the inability of IWC to have the required budget that enables it to provide its services in a sustainable manner. This led to the accumulation of sediment in the irrigation canals and deterioration of the water regulation structures, machinery and pumps. By the year 2000, the IWC was dissolved, and the MIWR is again responsible for the operation and maintenance (O&M) of the irrigation canals up to the minor off-takes. The Ministry of Finance and National Economy provides the MIWR with the annual budgets for operation and maintenance.

5.5. The Policy Regime in Water Quality Management

Apart from effluent regulations, and sometimes, national water quality guidelines, a common observation is that few developing countries (Sudan is not exception). Although water supply is seen as a national issue, pollution is mainly felt at, and dealt with, at the local level. With few exceptions, national governments in developing countries have little information on the relative importance of various types of pollution (agriculture, municipal, industrial, animal husbandry, and aquaculture), and therefore, have no notion of which is of greatest economic or public health significance (Salih, 1992). Usually, freshwater quality management is completely divorced from coastal management, even though they are intimately linked. Consequently, it is difficult to develop a holistic management plan for assuring water quality or to efficiently focus domestic and donor funds on priority issues. A Sudanese national water policy should include the following water quality components: (1) A policy framework that provides broad strategic and political directions for future water quality management. (2) A strategic action plan for water quality management based on priorities that reflect an understanding of economic and social costs of impaired water. This plan should include the following components:

- A mechanism for identifying national priorities for water quality that will guide domestic and donor investment.

- A plan for developing a focused and cost-effective data program for water quality and related uses, as a basis for economic and social planning.
- A consideration of options for financial sustainability including donor support, public-private sector partnerships, and regional self-support initiatives.
- A regulatory framework that includes a combination of appropriate water quality objectives (appropriate to that country and not necessarily based on Western standards) and effluent controls. This must include both surface and groundwater.
- A mechanism for public input into goals and priorities.
- A process for appointing specific agencies with responsibility for implementation so that accountability is firmly established and inter-agency competition is eliminated.
- Specific mechanisms for monitoring the quantity of drinking water, at the community level, if necessary.
- National data standards that must realistically reflect national needs and capabilities. The objective is to ensure reliable data from those organizations that provide information for national water management and for community-level drinking water quality.

The design criteria in any water quality program in Sudan should include the management issues for which water quality data are required. Generally, there are four categories of data objectives: (i) Descriptive data that are typically used for government policy and planning, meeting international obligations, and for the public information (Appendix 1). (ii) Data specific to public health. (iii) Regulatory mechanism and concerns, and (iv) Aquatic ecosystem health.

The last category is not normally included in many developing countries for reasons of cost and complexity (UNESCO, 1999). In most developing countries, countries with transitional economies, and some developed countries, the technology of monitoring has changed little since 1970s, yet some of the largest advances in monitoring in recent years involve technical innovations that serve to reduce costs and increase efficiency (Salih, 1992). Admittedly, not all are inexpensive; however when deployed appropriately, they may eliminate traditional monitoring or reduce costs by increasing the efficiency of more traditional approaches to chemical monitoring. Types of innovation include biological assessment, use of enzymatic indicators, miniaturization, automation, and simplification of laboratory analytical methods. Water quantity is highly variable in Sudan, reflecting regionally different levels of development, and is indirectly responsive to different needs for water quality programs (Tables 5). In Sudan, financial and sustainability issues include cost avoidance and cost reduction, local and accountability frameworks that encourage good business practices by senior program managers, the use of new cost-effective technologies for monitoring, and a variety of donor/public/private sector linkages that focus on commercial benefits that permit the transfer of certain parts of water quality programs to the private-sector.

5.6. Sustainable Development

In Sudan, there are a limited water resources and increased demands to cope with the rapid development in petroleum production. It is paramount to inaugurate strategies that control augmentation and conservation of water. Such measures essentially include uses of water, minimizing losses, quality protection, exploration, artificial recharge and water-harvesting techniques. A schematic technological advancement of low-cost water supply systems such as dug wells, roof top catchments, haffirs (in western and eastern of Sudan) and small dams combined with development of guidelines for settlement policy will hopefully lead to an improvement of water supply systems, water quality and reduction of the distance to the supply points. The following problems encountered so far need to be solved in the near future: (1) The establishment of efficient operation, maintenance and repair of water yards. (2) Community participation in operation and maintenance of water yards. (3) The extent to which capital costs can or should be recovered from water users.

Due to the difference in economic conditions in different areas, the development of rural energy is considerably imbalanced. The main challenge to energy policymakers in the 21st century is how to develop and manage adequate, affordable and reliable energy services in a sustainable manner to fuel social and economic development. Current rural energy relies mainly on charcoal, firewood and green energies, such as electricity and biogas. The bad economic situation leads to considerable difficulty in the development of rural energy, and farmers in remote areas still prefer “free firewood” for their cooking due to their low incomes. Hence, further development of rural energy needs significant financial support from the government at various levels. Although the work of rebuilding traditional stoves has been almost finished, most of the rebuilt fuel-saving stoves have a thermal efficiency less than 20%.

Table 5. The present water management of Sudan (NWC, 1994)

Use of resources	Sources	Administrative	Pricing Principle	Price Details
Urban	Surface & ground	(NWC)	Full cost recovery	Progressive rate with increasing uses. Rates lower in the north
Major rural villages	Mostly groundwater	(RWC)	Stand pipe free, recovery of recurrent costs, charges for yard and house connections	Progressive rates but less comparative to urban cities
Rural villages	Groundwater	(DC)	As above	Not available
Livestock	Surface & ground	(RWC)	All investments and recurrent costs	Regressive, no charges on relatively small use
Mines	Surface & ground	(NWC)	Full cost recovery	Progressive rates
Wildlife	Mostly surface	(RWC)	Full cost of boreholes	Regressive

National Water Corporation (NWC), 1994.
Rural Water Corporation (RWC), 1996.
District Councils (DC), 2000.

Chapter 6

SMALL WATER ENTERPRISES (SWEs)

Water has become an expensive commodity for the poor people, but it remains cheap for the rich. It is surprising that costs that are being recovered from the poor are used to subsidize cheaper supplies to the rich. Water is now a commodity for the privileged. The 6,000 liters supplied free often run out in 11 days for those with big families. When the water runs out, residents are forced to hire vans for up to $2.75 per person to bring water from wherever they can successfully beg for it. However, the age-old tradition of community compassion has unraveled, and many neighbors are less willing to share (Figure 3).

In the poor settlements, where connection to tap water does not exist, local water vendors play a central role in water distribution. Water vendors cover between 95 to 98% of water distribution to households, with the remaining 2 to 5% of households fetching water themselves (Figure 4). The total population served by water vendors in Greater Khartoum is estimated to be about 2,450,000 people (35% of the area's total population). Almost all water vendors use donkey carts as a means for water distribution.

Donkey cart operators play an effective role in the provision of water to households in the informal sector (Figure 5). The number of vendors in the capital is estimated at 30,000, with businesses covering 35% of the capital in the urban and semi-urban settlements. In the new informal settlements, water vending constitutes a good business for local entrepreneurs who invest in water carts. The average daily income for those who own the cart is estimated at SD700 per day and SD500 for those who rent carts, while for those who practice profit sharing with cart owners, it is SD350, too little to depend on for livelihood.

Figure 3. Water Distribution System in Eastern Regions.

Figure 4. For some people, water collection is a daily need.

6.1. PROBLEMS

The following summarizes some problems that occurred:

- The water carts are made of low-quality materials, and they are poorly designed. The donkey harness is poor and may cause injury to the animal.
- The health section within the local council is supposed to monitor the hygiene situation of the water yards, but they do not provide an effective service.
- All vendors in the surveyed locations complained about the high fees for obtaining licenses, health certificates and work permits. It seems that the local council is interested in vendors as a source of revenue rather than important actors in water distribution to households.

Figure 5. A typical donkey-drawn water tank used by water vendors.

Khartoum State's water supply system:

- six water treatment plants
- 884 boreholes and water yards
- 612 hand pumps
- two small dams
- 127 local haffirs (water reservoirs)
- Over 200 open wells

Chapter 7

SAFE WATER FOR NEW MILLENNIUM

In the past decade, sustainability has increasingly become a key concept and ultimate global criterion for socio-economic development in the modern world. Without a doubt, sustainable development and management of natural resources fundamentally controls the survival and welfare of human society. Water is an indispensable component and resource for life, and essentially all human activities rely on water in a direct or in direct way. Yet, supplying water of sufficient quantity and safe quality has seldom been an easy task. Although sustainability is still a loosely defined and evolving concept, researchers and policy-makers have made tremendous efforts to develop a working paradigm and measurement system for applying this concept in the exploitation, utilization and management of various natural resources. For water resources, recent development has been synthesized and presented in two important documents published by ASCE, (1998), and UNESCO, (1999). Those organizations attempt to give a specific definition and a set of criteria for sustainable water use. When considering the long-term future as well as the present, sustainability is concept and goal that can only be specified and implemented over a range of spatial scales, of which urban water supply is a local problem with great reliance on the characteristics and availability of regional water resources.

Sudan needs assistance in developing and implementing (a) river-basin management, (b) diffuse source pollution, (c) environmental restoration, and (d) urban storm drainage. At present, the international, bilateral donor agencies, and relevant United Nations bodies provide such assistance, e.g., UNICEF, UNDP, and WHO. The international associations constitute an additional, but as yet untapped, source of assistance. Advanced research and technology contribute to resolving water shortage and sanitation problems, and non-conventional, and reliable water supplies cannot be provided unless environmental impacts are considered. Looking to the future, Sudan has set the following priorities for water-resource research and development until the year 2020 (Amir, 1998): (i) Maximize overall water-use efficiency. This could be achieved by (a) improving irrigation and assuring its flexibility to cope with modern farming (b) drawing up a proper mechanism for water pricing; (ii) Modify the cropping pattern; for example (a) planning different cropping patterns according to water quality, (b) gradually replacing sugar cane by beet, (c) introducing and developing of salt tolerance crops, and (d) reducing the area of clover (Berseem); (iii) Re-use all the possible agricultural drainage water using proper technical means to deal with its quality, especially after implementing the irrigation development; (iv) Plan properly the re-use of sewage effluent after drawing up guidelines for its use; (v) Research agreements of losses of water and suggest conservation projects; (vi) Use and management of reservoirs and groundwater sources in the Nile valley in an integrated fashion, giving special consideration to drought conditions; (vii) Develop on a sustainable basis, sustainable groundwater resources in the deserts; (viii) Harvest rainfall in desert areas and make full use of torrential streams and flash floods; (ix) Use new economical technology of seawater desalination; (x) Raise public awareness about water resource scarcity and government management plans; and (xi) Consider laws to match with required development and existing scarcity; (xii) Establish efficient operation, maintenance and repair procedures; (xiii) Assure community participation in operation and maintenance; (xiv) Assess the extent to which initial government investment can or should be recovered from water uses; (xv) Aim to assure domestic potable water supply reaches at least 25 liters per day per person; (xvi) Aim to assure water should be available for ten livestock units at 450 liters per day (L/d); (xvii) Assure that potable water available within two kilometers of individual residences. The challenges of enhancing water supplies in the twenty-first century are as follows: (a) Drinking-water sources should be treated chemically; (b) Suitable toilet facilities should be provided along the main roads to minimize pollution; (c) Proper arrangements should be made for litter dumping and waste disposal; (d) Local people should be fully educated about environment matters and hygiene; (e) Previous damage should not be allowed to continue while planning for a balanced development in the future; (f) The concept of the ecosystem (involving education and interpretation of the natural environment) must be promoted; (g) The polluters must be introduced.

Chapter 8

ECONOMY, AGRICULTURE AND FOOD SECURITY

The backbone of Sudan's economy is its agricultural sector. The agricultural sector determines, to a great extent, the economic performance of the Sudanese economy. In fact, the country can be rescued by proper organization and utilization of its agricultural potential. Although endowed with rich natural resources, Sudan remains comparatively underdeveloped, primarily as a result of protracted civil strife and poor economic management. The economy showed a limited response to reform packages during the 1980s and early 1990s. Budget deficits have been common, the average annual rate of inflation peaked at 70% for the period of 1991 to 1995, but gradually subsided to less than 5% in 2001, then climbed to 8% in 2002. Interest rates remained negative during that period and resulted in the collapse of savings, affected the banking system adversely and eroded public confidence.

The GDP of the Sudan was US$17.8 billion (current US$) in 2003. The agricultural sector is the most dominant in the country's economy, even though its share has declined recently because of decreased agricultural production and the increased exploitation and export of mineral oil. In 2002, the sector contributed over 39% to the GDP and employed 57% of the total economically active population in 2004 (Appendix 3). It contributed about 90% of the Sudan's non-oil export earnings.

Sudan's agro-ecological zones support a variety of food, cash and industrial crops. Vast natural pastures and forests support large herds of livestock including cattle, sheep, camels and goats. The main exported crops are cotton, Arabic gum, sesame, groundnuts, fruits and vegetables; livestock is also important for exports. Within the agricultural sector, crop production accounts for 53% of agricultural output, livestock for 38% and forestry and fisheries for 9%.

Rain-fed agriculture covers, by far, the largest area in Sudan. The area actually cultivated and total yield may, however, vary considerably from year to year, depending on variability of rainfall. The rain-fed farming system is characterized by a small farm size, labor-intensive cultivation techniques employing hand tools, low input level and poor yields. Crops grown in the rain-fed sector include sorghum, millet, sesame, sunflower and groundnuts. According to the latest estimates, the traditional rain-fed farming sector contributes all the production of millet, 11% of sorghum, 48% of groundnuts and 28% of sesame production of the country. Mechanized, rain-fed agriculture comprises about 10,000 large farmers, with farm sizes of 400 to 850 ha, and a few large companies with holdings of 8,400 to 84,000 ha.

Sudan has the largest irrigated area in sub-Saharan Africa and the second largest in the whole of Africa, after Egypt. The irrigated sub-sector plays a very important role in the country's agricultural production. Although the irrigated area constitutes only about 11% of the total cultivated land in Sudan, it contributes more than half of the total volume of the agricultural production (Figure 6). Irrigated agriculture has become more and more important over the past few decades as a result of drought and rainfall variability and uncertainty. It remains a central option to boost the economy in general and increase the living standard of the majority of the population.

Sudan is generally self-sufficient in basic foods, albeit with important inter-annual and geographical variations, and with wide regional and household disparities in food security prevailing across the country. The high-risk areas are north Kordofan, north Darfur, the Red Sea, Butana and the fringes of the major irrigation schemes in addition to the Southern States. Major constraints to higher farm productivity and incomes are high marketing margins on agricultural produce and an inadequate allocation of budgetary resources and of the scarce foreign exchange earnings. As a result, the low-input/low-productivity model of production continues to prevail, and small farmers' incomes remain depressed. In the wake of the food shortages experienced in the 1980s, a high priority has been given by the government to producing food crops. This has resulted in large expansions in sorghum and wheat areas and output. Much of this has been at the expense of the main cash crop, cotton, with production declining by more than 40% since the mid-1980s.

8.1. Water Management

The Gezira Scheme is managed on a vertically integrated basis by the semi-autonomous Sudan Gezira Board (SGB). The MIWR is responsible for managing the Sennar Dam on the Blue Nile and the upper reaches of the irrigation system, responding to requests for water delivery from SGB's field staff. Within the scheme, the SGB serves as landlord, operates and maintains the lower reaches of the irrigation system and provides most of the inputs and services required by farmers to produce cotton, which is transported by the Board to its ginneries and sold on behalf of growers by the Sudan Cotton Company Limited. The SGB recovers the cost of advances made for inputs and services from the cotton sales before payment is made to the farmer. Tenants are wholly responsible for growing other crops in prescribed rotations with cotton (sorghum, groundnuts, forage, wheat, and vegetables), making their own arrangements for input supplies and marketing. By 2001, in the Gezira Scheme, Minor Canal Committees had been formed along the minor irrigation canals, and representatives of each of these committees constitute the Irrigation Committee at the block level. In addition to the Irrigation Committee, a Financial Committee has been established that is coordinating the reimbursement of the seasonal credits and arrangements for procurement of new inputs. The Irrigation Committee, with representatives of each of the minor canal committees, will be responsible for the operation and maintenance of the minor irrigation system, a task presently entrusted to the SGB, with the Ministry of Irrigation responsible of supplying the main system.

To address some of the problems facing irrigation management and development, the government has formalized a policy framework that includes:

- Transferring the operation and production of large- and medium-size irrigation schemes to the farmers and giving them full responsibility for water management on the irrigation system below the minor canals level through establishment of voluntary water users associations (WUAs).
- Fostering sustainable productivity of the large schemes through rehabilitation, combined with financial and institutional reform.
- Grouping, rehabilitating and handing over the relatively small-size pump schemes in the Blue Nile and the White Nile. These schemes were originally established and run by the government. Recently, and in accordance with the economic reforms, these schemes were handed

over to the private sector, represented by individual farmers, cooperatives or private companies.

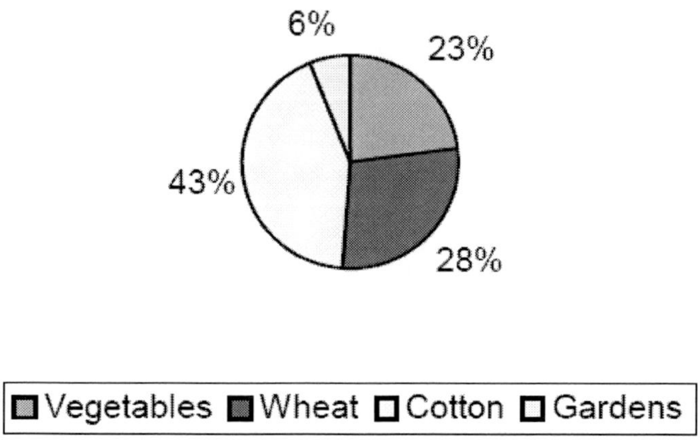

Figure 6. Irrigated plants.

8.2. Evolution of Irrigation Development

Irrigation potential was estimated at about 2.78 million ha, based on soil and water resources criteria. This figure does not take into account possible large-scale developments in the enormous wetlands in southern Sudan.

The Gezira Scheme is Sudan's oldest and largest gravity irrigation system, located between the Blue Nile and the White Nile. Started in 1925, and progressively expanded thereafter, it covers about 880,000 ha. It receives water from the Sennar Dam on the Blue Nile and is divided into some 114,000 tenancies. Farmers operate the scheme in partnership with the government and the Sudan Gezira Board, which provides administration, credit and marketing services. The scheme has played an important role in the economic development of Sudan, serving as a major source of foreign exchange earnings and of the government revenue. It has also contributed to national food security and in generating a livelihood for the 2.7 million people who now live in the command area of the scheme.

In the 1970s, Sudan was expected to become the "bread basket" of the Arab world, and with large investments from oil-rich Gulf nations, irrigation schemes such as the Rahad Scheme, which receives its water from the Rahad River and the Blue Nile, were established. Large-scale irrigated agriculture expanded from 1.17 million ha in 1956, to more than 1.68 million ha by 1977.

The 1980s were a period of rehabilitation, with efforts to improve the performance of the irrigation sub-sector. In the 1990s, some smaller schemes were licensed to the private-sector, while the four big schemes of Gezira and Managil, New Halfa and Rahad remained under the government control because they were considered strategic schemes.

In 2000, the total area equipped for irrigation was 1,863,000 ha, comprising 1,730,970 ha equipped for full or partial control irrigation and 132,030 ha equipped for spate irrigation. Only about 800,000 ha, or 43% of the total area, are actually irrigated, owing to deterioration of the irrigation and drainage infrastructures. In 1995, surface water was the water source for 96% of the total irrigated area land, and the remaining 4% were irrigated from groundwater (small tube-wells). The irrigated area where pumps are used to lift water was 346,680 ha in 2000. Most irrigation schemes are large-scale, and they are managed by parastatal organizations known as Agricultural Corporations, while small-scale schemes are owned and operated by individuals or cooperatives.

A number of water harvesting projects were implemented in the western part of Sudan during the 1970s, 1980s, and late 1990s. The main objective was to combat the effects of drought by improving crop production and increasing domestic water use. However, few of those projects have succeeded in combining technical efficiency with low cost and acceptability to the local agro-pastoralist farmers. This is partially owing to the lack of technical know-how, but also because of the selection of inappropriate approaches with regard to the prevailing socio-economic conditions.

8.3. Perspectives for Agricultural Water Management

The country has an agricultural potential of 105 million ha, of which only 16.7 million ha are cultivated and only about 1.9 million ha out of an irrigation potential of 2.8 million ha are under irrigation now. Therefore, there is ample room for further developments, especially in the irrigation sub-sector. However, there are three major constraints to irrigation development in Sudan:

- The ineffective process of annual maintenance of civil works (reportedly due to lack of funds), especially for the removal of silt and weeds from irrigation canals, which slows down water flow and causes continuous shrinkage of the actual irrigated area;

- The steady increase in development costs, which has been aggravated by the continuous devaluation of the local currency;
- The lack of farmer involvement in the planning and operation of the schemes and related services.

Generally, the water supply for all of the irrigation schemes is provided by dams and/or pumps and extensive networks of canals covering the whole schemes as well as drainage networks of canals. The overall objective of water management policies is to improve water use efficiency in agriculture, which includes efficient control of water in the irrigation networks, maintenance of the irrigation structures, provision of technical capacities capable to operate the systems, and efficient and economical maintenance of the irrigation systems. Supplementary irrigation could increase the very low or zero productivity of crops and fodder. The conjunctive use of groundwater and surface water could help to optimize the water resource productivity.

Chapter 9

POLLUTION FROM RIVER NILE

From a visual investigation of the River Nile (Table 6), the major sources of pollution are industrial effluents, crude sewage from blocked, broken or overloaded sewers, sewage effluents, surface runoff, and solid wastes that have been dumped into the river. Therefore, remedial and improvement measures must be taken before the environment becomes further polluted and the natural resources are completely over-exploited (Omer, 2000).

Table 6. Types and sources of wastes in River Nile water (Omer, 2000)

Materials sources	(%)
Paper and wood	50.0
Ferrous residues	12.5
Glass	11.0
Organic wastes	10.0
Plastics	5.0
Non-ferrous residues	1.5
Other	10.0

9.1. ENVIRONMENT AND HEALTH

The depositions of silt in irrigation canals and the subsequent built-up of aquatic weeds result in losses in production of up to 40%. Other costs of siltation include loss of hydropower potential since methods of silt removal involve measures that lower the head and interfere with generator operation. The most serious effect, however, is the loss of agricultural production. Agricultural chemicals were introduced into Sudan in 1946, in the Gezira scheme. Since then, the application of chemicals has intensified and proliferated into other agricultural schemes and in the private vegetable and horticulture fields as well as for control of desert locusts, birds and rodents and for public health. A total of about 200 active ingredients are registered in Sudan in over 600 different formulations of pesticides. An average of about 450 tons of insecticides and 150 tons of herbicides were applied annually during the period of 1993 to 1997. The annual consumption of fertilizers in the whole of Sudan is estimated at 80,000 to 200,000 tons of urea and 20,000 to 40,000 tons of super phosphate. The applied pesticides or the residue and degradation products can contaminate the water resources from the formulating sites, fallout from the spray, washing from contaminated clothes, empty containers, application equipment and dumping of the surplus. Serious contamination has been detected in the Gezira canals as well as in boreholes in the Qurashi area (Hassahessa Province) and the Kassala horticulture area. Fertilizers containing inorganic nitrogen as well as waste containing organic nitrogen are the two main sources of frequently reported nitrate and nitrite contamination in groundwater. Detailed studies are needed to discriminate between the two pollution sources.

9.2. Status and Evolution of Drainage Systems

Due to excess rainfall and sometimes to misuse of irrigation water, all the irrigation schemes need drainage networks to remove any excess water from the cultivated areas. In low areas, minor drains and collector drains are constructed to remove this excess water by gravity into cutout low areas or into natural drains. Sometimes, pumps are used to take water from low lands into areas outside the scheme. Also, escape drains are constructed along the main canal to carry any excess water to the nearest river or natural drain. It is estimated that in 2000, about 560,000 ha were drained. The development plans for the irrigation sector include the rehabilitation of the existing irrigation schemes, a shift of emphasis towards the development of small-scale irrigation schemes, and phased development and vertical expansion.

Chapter 10

THE CHALLENGE OF OVERCOMING THE COUNTRY'S DIVERSITY

Table (7) shows main water resource issues in Sudan by region. Proper water management is essential to sustain development. Competing for water are municipal supply, industry, and agriculture.

The National Water Act of 1994 (Law No. 1155) defines the objectives, principles, and instruments of the National Water Resources Policy and the National Water Resources Management system. The law establishes the institutional arrangement under which the country's water policies are to be implemented. The National Water Resources Policy was proposed to achieve: (1) Sustainability: to ensure that the present and future generations have an adequate availability of water with suitable quality. (2) Integrated management: to ensure the integration among uses in order to guarantee continuing development. (3) Security: to prevent and protect against critical events, either arising from natural causes or inappropriate uses. To achieve such objectives, water management must be implemented according to the following principles: (i) Water is a public good. (ii) Water is a finite resource that has economic value. (iii) The use of water required to meet people's basic needs shall have priority, especially in critical periods. (iv) Water management shall comprise and include multiple uses. (v) The river basin is the appropriate unit for water management, and (vi) Water management shall be decentralized, with participation by the government, stakeholders and society.

Table 7. Main water resource issues in Sudan by region (Omer, 2001)

Region	Water resource issues
South	1. Abundant water resources
	2. Localized scarcity of water and untapped water supplies
	3. High hydropower potential
	4. Water conflicts arising from immigration of Bagara Arabs (nomadic) from north to south
	5. Water-borne diseases
	6. International water conflicts (upstream and downstream countries)
Central	1. Water quality problems from untreated sewage and other pollution
	2. Water-borne diseases
	3. Potential use of rivers for navigation and recreational purposes
	4. Intensive erosion and sedimentation from agriculture
	5. High hydropower potential
	6. Excessive use in large urban and industrialized areas
	7. Frequent urban floods
North	1. Good water quality
	2. Scarcity of water resources
	3. Intensive erosion and sedimentation from agriculture
	4. Frequent urban floods
Northeast	1. Scarcity of water resources
	2. Water quality problems from untreated sewage
	3. In mining areas, water quality problems from effluent
West	1. Scarcity of water resources
	2. Water conflicts between nomadic and non-nomadic tribes
	3. Water-borne diseases
	4. Soil erosion and degradation caused by agriculture

10.1. Water Scarcity Impacts and Potential Conflicts

The failure of water resources to meet the basic requirements of society has a host of social, economic, environmental, and political impacts. Water scarcity is man-made phenomenon brought about by population pressures. The imbalance in the population-water resources equation strains society and has an adverse impact on domestic hygiene, public health, and cost of domestic water, and could impart political problems as a serious as bringing down government (Amir, 1998). On the social side, water scarcity adversely impacts job opportunities, farm incomes, credibility and reliability of agricultural exports, and ability of the poor to pay for domestic water. Economically, the adverse impact is displayed in the loss of production of goods, especially agricultural goods, and the loss of working hours because of the hardships society faces as a result of water scarcity. The impacts of water scarcity on regional stability are addressed with reference to water in the East Africa (Egypt and Sudan downstream and Ethiopia, Uganda, Kenya, etc., Upstream), taking into account the serious impacts of conflicts and potential water war (Amir, 1998). Conditions of scarcity propel an increase in competition among the different sectors of water use with results, invariably, at the expense of irrigated agriculture (Amir, 1998). Pure market forces create a pressure gradient under which water flows from the poor to the rich. Tough decisions await politicians, and the consequences are expected to displease some parties and please others. The scene of domestic politics becomes as fluid as water itself, with politicians shifting positions continuously in response to domestic pressures. The political fallout from water scarcity on the domestic scene is parallel to the impact the scarcity has on domestic households in terms of basic needs for drinking and food preparation, on domestic hygiene, and on public health. Other important factors have a delayed response to water scarcity, and these pertain to the integrity of the environment, and deterrence in development investment and economic credibility of the country. The cost of mitigating these problems and of the provision of services to increased urbanization could very well be beyond the ability of government to bear. The political consequences resulting from this will not be in favor of domestic stability, and social explosions can be anticipated (Omer, 2000). Water relations can be transformed into a positive sum games by which all parties can win. One common gain to all is environmental protection of the common watercourse or water body. Lack of cooperation and agreement generally leads to environmental neglect and water quality degradation, which is loss to all. International encouragement to attain cooperation can, therefore, be brought to bear on the regional parties, and efforts of international lending agencies can be called upon to pool with the regional and international efforts to achieve this objective (e.g., Great Lakes, Canada and USA).

Today's advanced societies heavily depend on energy. The sources of energy and electricity generation today in Sudan are waterfalls, natural and artificial, and fossil fuel. Energy from waterfalls is short of meeting the current or future energy requirements, and the fossil fuel resources, being depleted with time, will eventually run out. For human civilization to continue at its natural pace, new forms of affordable and clean energy must come on line. Failure of human civilization to introduce new forms of energy will render that civilization doomed, and the quality of life will deteriorate. If this unlikely scenario actually takes place, the requirements will decrease because the mechanism of making it available for use (pumping) diminishes. The more likely scenario is more optimistic one, and it is that a new form of energy generation will be introduced in which case water desalination becomes affordable and its pumping from the coastal desalination plants becomes possible at reasonable cost. The way out of the looming water crisis rests, therefore, in the invention of new forms of energy generation that will make possible the reliance on desalination and in the recycling of wastewater for reuse in agricultural production and for environmental reasons. Integrated management of the three resources of water, energy, and the environment will result in better results with a positive sum for society. It is well known that all existing sewage treatment plants are overloaded. Hence, the treated effluents do not comply with international effluent quality guideline. The main reasons behind this are

- Weak management and absence of environmental awareness,
- Public-sector institutional problems,
- Failure in process design, construction and operation,
- Lack of skilled operating staff and insufficient monitoring programs,
- Poor maintenance and weak financial resources, and
- Low level of public involvement and lack of financial commitment.

Chapter 11

PROSPECTS FOR WATER RESOURCE PLANNING IN NILE BASIN

11.1. COMMON LANGUAGE AND CULTURE

A common language and similar culture simplify communication and reduce the potential for misunderstandings. In the river Nile basin, where several languages are spoken, an international language, English, is used with some success by multi-jurisdictional basin management authorities.

11.2. FACTORS PROMOTING DATA AND INFORMATION EXCHANGE

Data and information exchange is more probable when needs are compatible and when there is potential for mutual benefit from cooperation (Table 8). Where countries are working on developments that are beneficial to both countries as well as other riparians, there is little incentive to hide project impacts. This means that because data and information exchange are unlikely to lead to pressure from surrounding countries that might restrict developments, countries have less reason to restrict access to their data and information resources. It is important for there to be no perceived clash of interests in development plans and needs. An example of this might be in developing their part of the basin primarily for hydroelectric development, while the lower riparians are more interested in developing the irrigation potential of their portion of the basin. By constructing large storage dams in the upper part of the basin, the river Nile seasonal flow might be evened out, reducing flooding downstream while increasing irrigation water supplies and even making downstream run-of-the-river hydroelectric projects more profitable.

11.3. Sufficient Levels of Economic Development

Sufficient levels of economic development across a basin are needed to permit joint funding of cooperative processes, particularly data collection and dissemination. Although countries with differing levels and forms of economic development may, at times, have more complementary needs than countries with similarly structured economies, the overall level of economic development is still significant. A wealthier country in a river basin may be able to assist with the funding of data collection activities in the neighboring country with much-needed data and helping to build confidence between the two countries.

Table 8. Summary of the situation relating to data and information exchange in Nile basin (H & D, 2003)

River basin	Nile basin
Basin states or territories	Burundi, Democratic Republic of Congo, Egypt, Eritrea, Ethiopia, Kenya, Rwanda, Sudan, Tanzania, Uganda
Cooperative frameworks in place	Nine of the countries of basin are pursuing the development of a cooperative framework
Major languages spoken in	More than six official languages and numerous unofficial languages
Major water issue facing the basin	Rapid population growth, environmental degradation, under development
External funding of cooperative basin initiatives	Extensive external funding of cooperative initiative
Range of GDP *per capita* of the basin	\$550-\$3000
Extent of data/information exchange	Information exchange through the cooperative framework being developed is beginning to occur

11.4. Increasing Water Resources Stress

There was a bilateral agreement between Egypt and the Sudan in 1959, by which the two countries share the Nile flow: 55.5 billion cubic meters to Egypt, 18.5 billion to Sudan, and 10 billion were allocated to evaporation. Hopes are high for achieving a more extensive participation by the other riparian parties in what could be a multilateral treaty on the Nile encompassing the other riparian states in addition to Egypt and Sudan. The above agreement is incomplete; lack of the entry of other legitimate riparian states, lack of the water quality components, and the focus on quantity and misses important management issues. It is to be noted that regional relations, including those among the riparian parties, are not disconnected from the political, economic, and trade network of international relations. Water is not the only determinant factor in shaping the nature of bilateral, regional, or international relations. As *per capita* water resources availability decreases (Table 9), tensions between riparian nations may rise and make cooperation difficult. Stress may, therefore, reduce cooperation and data sharing rather than strife. The historical background of the basin may have a lasting effect on current negotiations. Past conflicts can have a deleterious effect on the prospects for establishing cooperative practices, such as data sharing. Where there is a history of conflict between two nations, both nations may view the present situation primarily as competitive and focus on conflicting rather than common interests. Democracies may find it easier to negotiate cooperative arrangements with other democracies. Political differences can lead to legacies of mistrust developing between countries.

Table 9. Diverse water challenge

Country	Egypt	Sudan
Per capita annual water resources 2000 (m^3)	34	1187
Per capita annual withdrawal (m^3)	921	666
Per capita annual withdrawal for agriculture (m^3)	86	94

Source: World Resources Institute 1998 and 2000.
World Resources 1998-1999.
World Resources 2000-2001.

Climatic and environmental changes and a rising demand have increased the competition over water resources and have made cooperation among countries that share a transboundary river an important issue in water resources management and hydro-politics. Yet in river basins around the world, international conflict and cooperation are influenced by different factors, and general conclusions about forces driving conflict and cooperation have been difficult to draw. Rivers are an essential natural resource closely linked to country's well-being and economic success. But rivers ignore political boundaries, and competition over the water resources has lead to political tension between countries with Transboundary Rivers. Integrating international cooperation and conflict resolution into the water management of Transboundary Rivers has, therefore, become an important issue in water resources management and hydro-politics. The problem requires a good understanding of the history and patterns of conflict and cooperation among nations sharing international basins worldwide and of the different factors that have influenced their international relations. Increasing water scarcity in the downstream areas of several river basins demands improved water management and conservation in the upper reaches. Improved management is impossible without proper monitoring at various levels.

Chapter 12

DISCUSSIONS

Water is one of the most precious natural resources on earth. It is essential for human survival and development and cannot be replaced by any other resource. However, with rapid social and economic development, as well as explosive population growth, a water crisis has developed in Sudan during the 20th century. With economic development and population growth, the conflict between water supply and demand has become more and more acute in Sudan, and it has been aggravated further by the irrational utilization of water resources. As a result, the deterioration and destruction of the eco-environment have become increasingly serious. In order to effectively protect ecosystems and improve their ecological conditions, many developments on ecological and environmental water requirements have been carried out, including rivers, vegetation, lakes, wetlands and groundwater. Changes in the economy or in the population will have different impacts on the water resource availability, which in turn will impact economic output and population dynamics. Water conservation is a major challenge because of increasing competition between agricultural and non-agricultural use of water. Efficient use of water in agriculture is critical because of the large volumes of water used.

According to the 1993 census, 65 percent of the urban population have access to piped water (UNICEF, 1996), whereas only 20 of the rural population enjoy the same facility (Table 10). Other sources give much lower figures. The Government National Program of Action for Child Survival and Development sets availability of 20 liters of safe water per capita per day, within one kilometer from the users' dwelling, as the acceptable standard. A clear national definition of what a safe source of water is has yet to be adopted.

Urban populations theoretically consume 20 liters per capita per day, while rural people get eight liters at two and a half times the cost per liter. Variations are large from one place to another. People in Darfur consume less than one-fourth of those in Khartoum. Population in pre-urban areas and in newly emerging urban centers are around 35 percent of the total urban population and generally do not have access to safe water. The quality of urban water supply is questionable (Omer, 2004). Surface water sources are contaminated and bear heavy loads of suspended silt. Waterworks are old and small and cannot cope with the escalating demand. Increased reliance on ground water in the past few years has partially released pressure off in the way of quantity but not quality. Septic tanks and soak-away wells are a constant threat of contamination.

Distribution networks are old and a potential weak link taking up contaminants, in the case of the capital Khartoum, for example. Increased reliance on booster pumps and storage tanks have contributed to the depletion of the major power supply as well as offering excellent breeding sites for mosquitoes. The worst-case scenario is the city of Port Sudan, suffering from chronic water shortages on the one hand and the heavy incidence of malaria on the other.

The celebrations of the city of El Obeid with the new source of water from Bara, was short lived. The old distribution system soon collapsed. The storm drain is another fiasco. It is a paradox that there is no clear link between the water supply and sanitation sectors. Even within the water sector, there are poor linkages between the federal and state water corporations. The water sector itself has experienced ten major institutional changes since independence.

Table 10. Percent population with access to sources of water

Sources of water	Northern States	Southern States	Rural	Urban
Pipe connection	19.62	65.81	35.33	7.02
Boreholes	56.12	20.25	43.93	55.16
River/canal	14.09	2.25	10.05	30.01
Tanker	4.78	0.59	3.35	0.24
Haffirs/Fola	1.36	7.17	3.34	5.40
Others	3.97	3.82	3.92	1.74
Not stated	0.06	0.11	0.08	0.43
Total	100.0	100.0	100.0	100.0

The Sudanese National Plan of Action (NPA) sets the goal of universal access to safe drinking water and sanitary means of excreta disposal by the turn of the decade (Omer, 2004). The Comprehensive National Strategy gives priority to the following strategies for achieving the set:

- Cost-effective utilization and management of water resources.
- Introduction of low-cost appropriate technologies and encouragement of local production of equipment.
- Rehabilitation of deteriorating water sources and systems.
- An expanded program of well-drilling and hand-pump installation, especially in priority rural areas.
- Training, capacity building and increased use of domestic technical resources, to increase cost efficiency and reduce dependence on external resources.
- Development and expansion of sanitation services.
- Increased community involvement in planning, execution and management of water supply and sanitation services encouraging cost sharing and self-help.
- Encouragement of research aimed at better water resources management, evaluation of existing schemes and identification of cost-effective alternative strategic elements.

The National Comprehensive Strategy gives priority to the rural sector, emphasizing rehabilitation of existing water yards, expansion of low-cost technology options (hand pumps) and improvement of surface water sources (water harvesting, haffirs, etc.). The priority in urban areas is to restore existing waterworks to their original designed output, followed by rehabilitation of other sources and expansion of services, especially in Khartoum and the sixteen state capitals recently upgraded from towns to cities (Table 11). Only 40 percent of the urban population have access to safe water. Investment does not match growth in needs. The State Ministry of Housing and Public Works of Khartoum conceded, "The estimated deficit of clean water supply in Khartoum is about 40 thousand cubic meters per day." The unaccounted for losses in the distribution system are as high as 40% (Table 12).

Globally and regionally, a realization is setting in that constraint of freshwater resource availability and their sharing could be a major impediment to security and subsequently to sustainable development of developing countries. Scarcity of freshwater resources globally is fast becoming a major factor in limiting the development of regions and in causing conflict among regions sharing the same water resource. The Nile has many origins, draining ten percent of the African continent's river water and flows 6825 km through three climatic zones and ten sovereign countries. The Nile flowing through Sudan is the confluence of two major rivers. The White Nile and the Blue Nile merge near Khartoum, the Sudan capital.

Table 11. Rural water sources

Type	Number	Capacity (m^3/ day)	Installed actual yield
Water yards	4,252	447,000	146,000
Hands pumps	8,000	48,000	32,000
Dug wells	7,000	56,000	44,800
Haffirs/ Dams	856	142,000	56,800
Slow sand filters	233	16,500	8,200
Total	20,441	710,000	287,800

Sources: UNICEF, 1996.

Table 12. Urban Demand for water

Region	Capacity m^3/day	Demand/m^3/day	Coverage (%)
Khartoum	350,000	700,000	50
Northern	30,000	64,000	47
Eastern	48,000	205,000	20
Central	80,000	184,000	39
Kordofan	23,000	114,000	17
Darfur	14,000	74,000	20
Southern	20,000	103,000	16
Total	565,000	1,445,400	39

Sources: UNICEF, 1996.

Table 13. Water consumption in different African countries

Country	Population (10^6)	Total mean precipitation (km^3)	Sectors (%)	Per capita consumption (liters/day)
Botswana	1.7	261.9	Domestic 25% Agriculture 37% Industry 25% Others 13%	221.5
Burundi	7.1	NA	Domestic 58.3% Agriculture 21.8% Industry 5.2% Others 14.7%	260.7
Congo	3.6	NA	NA	111.1
Egypt	73	80	NA	200
Eritrea	4	NA	Domestic 36.9% Industry 34.8% Others 26.9%	NA
Ethiopia	67.8	NA	NA	12.25
Kenya	33	NA	Domestic 27.6% Industry 33.2% Others 39.2%	143
Libya	5.3	43	Domestic 11% Agriculture 86% Industry 2% Others 1%	100
Rwanda	7.9	80	NA	43
Sudan	39	80	Domestic 20% Agriculture 70% Industry 5% Others 5%	20

Reproduced from (H & D, 2006).

Table 14 Water supply norms: the minimum required for human consumption (WHO, 2006)

Purpose	Stand posts / hand pump supply	House service connections
Drinking	3	3
Cooking	5	5
Bathing	15	20
Washing utensils and house	7	12
Ablution	10	15
Washing Clothes	Nil	Nil
Flushing WC	Nil	15
TOTAL	40	70

The water, collected from surface or ground water sources, is conveyed to treatment plant for removal of impurities. The treated water, after adding quality control contents, is stored in clean water reservoirs and pumped to overhead tanks directly into the distribution system for supplying the customers through individual connection (domestic/non-domestic/industrial). In addition, street taps are provided for domestic consumption.

While implementing the rural water supply schemes, the following norms are adopted for providing potable drinking water to the population (Table 14):

Forty liters per capita per day (LPCD) for humans to meet the following requirements:

Purpose	Quantity (LPCD)
Drinking	3
Cooking	5
Bathing	15
Washing utensils and house	7
Ablution	10

12.1. Water Stress in Sudan

Water stress refers to economic, social, or environmental problems caused by unmet water needs. Lack of supply is often caused by contamination, drought, or a disruption in distribution. In an extreme example, when Sudan split four years ago between the rebel-led west and the government-ruled north, the conflict led to unpaid water bills, which precipitated a dangerous health threat in the region, increasing the risk of water-borne diseases such as cholera. Some analysts believe the disruption of distribution was a political ploy to put pressure on the rebel-led west.

While water stress occurs throughout the world, no region has been more afflicted than sub-Saharan Africa. The crisis in Darfur stems in part from disputes over water: The conflict that led to the crisis arose from tensions between nomadic farming groups who were competing for water and grazing land—both increasingly scarce due to the expanding Sahara Desert. As Mark Giordano of the International Water Management Institute in Colombo Sri Lanka says, "Most water extracted for development in sub-Saharan Africa-drinking water, livestock watering, irrigation—is at least in some sense 'transboundary'." Because water sources are often cross-border, conflict emerges.

Improving water and sanitation programs is crucial to spurring growth and sustaining economic development. Because it takes time to develop these programs, a paradox emerges: poor economies are unable to develop because of water stress, and economic instability prohibits the development of programs to abate water stress. Developments in water storage could have prevented that drought from significantly affecting Sudan's economy. Hydropower can also spark economic development. Accordingly, some transboundary water agreements also play a clear role in fostering development, for example, by facilitating investment in hydropower and irrigation.

12.2. The Role of Agriculture in Water Stress

Agricultural development has the potential to improve African economies but requires extensive water supplies. These statics from the Water Systems Analysis Group at the Institute for the Study of Earth, Oceans, and Space at the University of New Hampshire reveal the urgent need for sustainable agricultural development:

- About 64 percent of Africans rely on water that is limited and highly variable;
- Croplands inhabit the driest regions of Africa, where some 40 percent of the irrigated land is unsustainable;
- Roughly 25 percent of Africa's population suffers from water stress;
- Nearly 13 percent of the population in Africa experiences drought-related stress once each generation.

Another aspect of water-related stress is the relationship among water, soil, and agriculture. Improved access to quality water is a long-term goal that requires more than humanitarian funds:

- Because sub-Saharan Africa is subject to more extreme climate variability than other regions, it needs improved water storage capacity. Some experts say that large dam projects would create a more sustainable reserve of water resources to combat the burden of climate fluctuations; but others disagree, stating the harmful environmental impact of large dams.

- Many experts say more water treaties are needed. The transboundary water agreements have cultivated international cooperation and reduced the "probability of conflict and its intensity."
- Better donor emphasis on water development is needed. Small-scale agricultural improvements also offer a solution to water stress, including the harvest of water in shallow wells, drip irrigation for crops, the use of pumps, and other technological innovations.

Farmers can access green water through drip irrigation systems that slowly and consistently deliver water to plant's root system, supplemental irrigation (supplementary to natural rainfall rather than the primary source of moisture during periods of drought) and rainwater harvesting (the collection of rainwater for crops, which reduces reliance on irrigation). Crops can grow poorly even during periods of rainfall, and most farms in Africa suffer from nitrogen and phosphorus depletion in soil. One way to assuage water stress in terms of food scarcity is to increase water-holding capacity with organic fertilizers that would increase availability and efficacy of green water.

12.3. Water Supply Problems in the Butana Region-Central Sudan with Special Emphasis on Jebel Qeili Area

The Butana region of central Sudan is famous for its animal wealth and extensive pastures. Yet scarcity of water resources in the area, especially during the dry seasons, handicaps the proper utilization of these pastures. The area is occupied by non-water-bearing basement rocks and the only source of water is from direct run-off. Thus large numbers of small-size water reservoirs, "haffirs," were constructed, but these are inadequate to provide enough water for the growing human and animal population. An all-year lake is here proposed to be constructed utilizing the ring-structure the Jebel Qeili igneous complex, central Butana. This lake is expected to solve the present water problem and meet the future demand of central Butana at the present rate of human and animal growth.

12.4. SOUTHERN SUDAN

World Vision began its work in Sudan in 1972, through a partnership with the African Committee for Rehabilitation of the Southern Sudan (ACROSS) to provide emergency relief aid to war-effected families. Efforts included the reconstruction of the Rumbek community hospital and surrounding buildings, the provision of medicine and supplies, and education in preventative health care. Other projects during this period focused on training health and social workers in general medical aid and child welfare and instruction in water development, agriculture, handcrafts, and literacy.

The 1980s brought constant turmoil to the Sudanese people as the civil war raged on and severe drought parched the country. In 1983, approximately 1,500 refugees entered Sudan daily from violence-torn neighboring countries, straining the already limited food supply. World Vision, through the ACROSS Refugee Settlement Project, responded by distributing blankets, grain, cooking oil, medical kits, and shelter to more than 50,000 people. Supplemental feeding for children also was provided. Numerous development projects were initiated during this time that assisted communities in improved crop production, animal husbandry, health care, clean water collection, infrastructure repair, and literacy. In 1989, World Vision became a founding member of Operation Lifeline Sudan (OLS), a partnership of non-governmental organizations (NGOs) and the UN agencies designated to coordinate the southern relief efforts.

During the 1990s, World Vision conducted operations in all major regions of southern Sudan. Project objectives included primary health care, water provision, agriculture, local grain purchase, enterprise development, and emergency relief efforts. World Vision focused on an integrated work approach that involved peace and advocacy, gender development, church support, and environment and natural resource initiatives. Some specific projects included:

- The Kapoeta Medical Supplies Project provided health and educational assistance to more than 200,000 people to help reduce incidences of disease and suffering.
- The Agriculture/Livestock Rehabilitation Project assisted the aforementioned families with food, seed, vaccinations, and agricultural consults.

- The South Sudan Relief/Church Support Project coordinated a pastors' conference for 150 pastors and religious leaders from the Western Equatoria province.
- The South Sudan Relief and Rehabilitation Project, a ten-year program, provided 450,000 Sudanese with agriculture and economic development, food and water, healthcare, enterprise development opportunities, and emergency relief.

12.5. Beja People's Problems

The Beja, a semi-nomadic group of people who live in rebel-held areas of eastern Sudan, need a huge amount of humanitarian assistance, according to a representative from the International Rescue Committee (IRC). Although Beja can be found throughout northeast Africa, tens of thousands are currently trapped in an area of eastern Sudan near the Eritrean border, held by Sudanese rebels since the late 1990s. Only two NGOs, both based in Eritrea, are able to access the 15,000 sq km area at the moment, one of which is the IRC. The organization estimates the Beja population in the area to be between 45,000 and 186,000 people.

Although it did rain in the area in 2004, a shortage of water had also posed serious problems. "Fresh drinking water is incredibly hard to come by. All the settlements have just focused around dry river beds, in which people dig hand-dug wells." Locusts would eat the foliage that usually sustains the Beja's goats and camels—upon which the Beja utterly depend for survival. A few immature locust swarms have formed in northeast Sudan near the Red Sea and the border of Egypt, the UN Food and Agriculture Organization said in March 2004. Moreover, Beja grazing areas have been severely restricted by a front line between rebel forces and Khartoum government soldiers, the second to be opened by southern rebels during Sudan's 21-year-old civil war.

Sudan is an example that projects the environmental plight of Africa, south of the Sahara—drought and desertification, floods, deforestation, loss of biodiversity, tribal and ethnic conflict and poverty are only too common. As a result, interest and commitment to environmental impact assessment practices have become mandatory by donors when executing new development projects. The ecological zones of Sudan in 1998 as:

- Deserts: cover almost 30% of the northern parts. Annual precipitation is less than 50 mm; soils are sandy. Sparse vegetation grows on seasonal "Wadis" and the banks of the Nile.
- Semi deserts: cover above 20% south of the desert belt. Rainfall ranges from 50 to 300 mm. It is speckled with few Acacia trees and thorny bushes and zerophytes.
- Low rainfall woodland Savannah: covers about 27% of the area of Sudan with rainfall less than 900 mm, with a nine-month dry period. Annual grasses are dominant. Heavy clay soils lie on the east of the Nile, and the west is sandy. Most of the 36 million feddans of rain-fed agriculture and the four million irrigated lands fall within this heavily populated belt.
- High rainfall woodland Savannah: 13% of the area with rainfall more than 900 mm and with broad-leafed trees in the southern parts of Sudan.
- Swamps: are probably the largest in the world and cover about l0% of the area and fall in three main areas around the tributaries of the White Nile.
- Highlands: are less than 0.3% of the areas of Sudan and are scattered along the Red Sea coast, the south and the west of the country.
- The Red Sea Coast-Marine ecosystem, mangrove swamps, coral reefs and associated fauna.

Environmental problems include

- Horizontal expansion in rain-fed and irrigated agriculture;
- The complete absence of the environmental dimensions in policies, strategies, plans and programs of management of resources;
- Development is random and environmental evaluation does exist before or after execution of projects;
- The economy and society, in spite of the century-long attempts at "modernization" are still dominated by subsistence way of living;
- The economy is still affected seriously by the yearly, seasonal and geographical variability of rainfall for crop and livestock production;
- Dependence on imported seeds and agricultural chemicals has increased cost of production;
- Loss of land productivity and marketing policies decreased cash surplus;
- The civil war in the south has grave economic and social costs;

- Population distribution and rural-urban migration due to desertification and civil strife has led to deterioration of natural resources, indigenous knowledge and loss of local culture and dignity;
- Problems of poor sanitation, limited industrial pollution and food hygiene have become more complex;
- The energy crisis is aggravating desertification and affecting climate charge;
- Vast water resources are badly managed;
- Environmental education has only been recently incorporated in school curricula; and
- Laws and legislation concerning the environment are not effective, and law enforcement measures are not integrated.

12.6. Western Sudan

El Fasher, Darfur region, Sudan, 24 August 2005—Torrential rains have caused severe flooding in this city of 400,000 people and in nearby Abu Shook, a camp for people forced to flee their homes as a result of the ongoing Darfur conflict. The floods have destroyed hundreds of homes and have made El Fasher's water supply largely unsafe.

UNICEF is mounting a concerted effort to restore basic services to those affected by the flood, and to prevent the outbreak of disease. Since the flood, UNICEF has assisted with the following:

- Reinstalling pipes in Abu Shook and restoring the water supply by linking boreholes with pumps.
- Testing the water quality each day. No bacterial contamination has been found.
- Rebuilding 156 latrines and 88 bath stations.
- Renting five tankers to deliver more water.
- Repairing damaged schools and child-friendly spaces.
- Providing daily door-to-door hygiene-promotion trainings.
- Distributing jerry cans, soap, tarps, and mosquito nets.

Chapter 13

CONCLUSIONS

1. Booming economy, high population, land-locked location, vast area, remotely separated rural areas, which are not easily accessible, large reserves of oil, excellent sunshine, large mining sector and cattle farming on a large-scale are factors that are most influential to the total water scene in Sudan.
2. Water resources plans are developed to guide future decisions and are to be developed for each river basin and state, as well as for the country. The overall objective is to coordinate efforts and establish guidelines and priorities for water allocation and water pricing. The priorities established for water allocation would be used in critical drought conditions. The water quality classification of water bodies by different classes of use is the basis for truly integrating the quality and quality of water management. Water pricing is the single most controversial instrument of the law. The pricing system recognizes the economic value of water, as stated in the principles of the policy, but is also the most difficult step to implement. It is expected that the pace of implementation will increase and the quality of work will improve in addition to building the capacity of the private and district staff in contracting procedures. The financial accountability is also easier and more transparent.
3. The communities should be fully utilized in any attempts to promote the local management of water supply and sanitation systems. A change in water and sanitation sector approach from supply-driven approach to demand-responsive approach call for full community participation. The community should be defined in terms of their

primary role as user/clients. Private-sector services are necessary because there are gaps, which exist as a result of the government not being able to provide water services due to limited financial resources and increase in population. There is little notion of "service, invoice and move on." As a result, there are major problems looming with sustainability of completed projects.

4. The factors affecting the eco-environmental changes are complex. There are interrelated and interact. The deterioration problems of water and sanitation have attracted some attention in recent years. There is an urgent need to study possible rehabilitation measures to ensure a sustainable and excellent water quality and improved sanitation.

References

2006 World Atlas and Industry Guide. (2006). Hydropower and sustainable water resources management in the world. *The International Journal on Hydropower and Dams*, United Kingdom.

Amir, B. (1998). *Water development in Sudan.* Arab League. Khartoum: Sudan.

Anon. (1979). Map of the world distribution regions. *MAB Tech. Note* 7.

ASCE Task Committee on sustainability criteria. (1998). Sustainability criteria for water resource systems. Reston, Virginia, USA: ASCE.

Brown, L., and Halweil, B. (1999). *Vital signs.* 1999. New York, USA.

FAO United Nations Food and Agriculture Organisation. (1999). *The state of food in security in the world.* Rome: Italy.

Habeballa, H. (1991). The Treatability of the Blue and White Nile Waters for Drinking Purposes in Khartoum, Institute of Environmental Studies, University of Khartoum.

Hydropower and Dams (H & D). (2003). *World Atlas and Industry Guide 2003*. UK.

James, W. (1994). Managing water as economic resources. Overseas Development Institute (ODI). London: UK.

National Water Corporation (NWC). (1994). *Water resources in Sudan.* Khartoum: Sudan.

Noureddine, R.M. (1997). Conservation planning and management of limited water resources in arid and semi-arid areas. In: *Proceedings of the 9th Session of the Regional Commission on Land and Water Use in the Near East.* Rabat: Morocco.

Omer, A.M. (1995). Water resources in Sudan. *NETWAS* 2(7).

Omer, A.M. (2000). Water and environment in Sudan: the challenges of the new millennium. *NETWAS* 7(2): 1-3.

Omer, A.M. (2001). Water development in Sudan: Present and future challenges. Arab Organisation for Agriculture Development (AOAD).

Arabic Journal of Irrigation Water Management, Vol.2, p.48-58, Khartoum: Sudan.

Omer, A.M. (2004). Water resources development and management in the Republic of the Sudan. *Water and Energy International*. 61(4): 27-39.

Overseas Development Administration (ODA). (1987). Sudan profile of agricultural potential. Survey: UK.

Salih, A.M.A.; and Ali, A.A.G. (1992). Water scarcity and sustainable development. *Nature and Resources*. 28(1).

Seckler, D. (1992). Private sector irrigation in Africa-Water resources and irrigation policy studies. Winrock International Institute for Agricultural Development.

UNESCO Working Group M.IV. (1999). *Sustainability criteria for water resource systems.* Cambridge, United Kingdom: Cambridge University Press.

UNICEF. (1996). *Situation Analysis of Children and Women in the Sudan.* Khartoum: Sudan.

WHO. (2006). Water norms and attitudes. Geneva: Switzerland.

APPENDICES

APPENDIX 1. THE DESCRIPTIVE FORMULAS

Main criteria	Sub-criteria	Formulaes
Food	Grain security	$\sum_{It=1}^{N} \frac{Grain\ production\,(t/y)}{\mathrm{Re}\,quired\ grain\ consumption\ (t/y)}$
Economic	Value added	$\sum_{i=1}^{N}$ [Return (SD) – Cost (SD)] Where N = number of sectors (agriculture, industrial, domestic water supply, navigation, power generation)
Water	Water security	Water security = [total available water resources – Total water requirements] (10^6 m^3 / year)
	Water sustainability	$\sum_{i=}^{K} \frac{Volume\ of\ water\ used\ from\ resource}{Potential\ of\ resource}$ Where K = number of water resources (surface water, shallow groundwater, deep groundwater, desalination, etc.)
Socio-economic	National employment	$\sum_{i=1}^{K}$ Total employment of sector $_I$ Where K = number of sectors (agriculture, industrial, domestic water supply, power generation and navigation)
	New settlers	Total homes tender of agriculture sector
	Land per capita	$\frac{Total\ cultivated\ land}{Total\ population}\ (hectare\ /\ capita)$

Environmental	Water per capita	$\frac{Available\ water\ resources}{Total\ population}$ (m^3/(capita.y)
	Shisto parasites prevalence	Shisto parasites prevalence = number of shisto patients over the country
	Typhoid parasites prevalence	Number of typhoid patients over the country
	Agro-chemical *per capita*	$\sum_{i=1}^{N} \frac{Pesticide\ used\ in\ (kg/y)}{Population}$ Where N = number of pesticide used all over the country
	Industrial effluent integrated water-quality index	Integrated water-quality index for sewage or industrial wastewater PTi = $\frac{APLPi}{SP/Pi}$ IWQI = $\frac{1}{K}\sqrt{\sum_{i=1}^{i=K} PI_i^{\ 2}}$ Where: PTi = population index for parameter K APLPi = actual population loads of parameter K SPLPi = standard pollution load of parameter K $IWQI_{S/I}$ = Integrated water-quality index (S= Sewage wastewater, I= industrial wastewater) K = number of proposed water-quality parameters
	Domestic effluent integrated quality index	Same as above

APPENDIX 2. FACTS ABOUT SUDAN

Full country name	Republic of the Sudan.
Total area	One million square miles (2.5 x 10^6 square kilometers). Land 2.376 x 10^6 square kilometers.
Population	39 x 10^6 inhabitants (July 2006 est.).
Capital city	Khartoum (population 5 million).
Language	Arabic (official), English, Nubian, Ta Bedawie, diverse dialects of Nilotic, Nilo – Hamitic, Sudanic languages.
Religions	Sunni Muslim 70% (in north), indigenous beliefs 25%, Christian 5% (mostly in south and Khartoum).
GDP per head	US $ 533.
Annual growth	4% (1997 est.).
Inflation	23% (1998 est.).
Ethnic groups	Black 52%, Arab 39%, Beja 6%, Foreigners 2%, others 1%.
Agricultures	Agriculture is the backbone of economic and social development. 62% of the populations are employed in agriculture. Agriculture contributes 33% of the gross national products (GNP), and 95% of all earnings.
Animal wealthy	35 x 10^6 head of cattle. 35 x 10^6 head of sheep. 35 x 10^6 head of goats. 3 x 10^6 head of camels. 0.6 x 10^6 head of horses and donkeys. Fish wealth 0.2 x 10^6 tons of food annually. Wildlife, birds and reptiles.
Environment	Inadequate supplies of potable water, wildlife populations threatened by excessive hunting, soil erosion, and desertification.
International agreements	Party to: Biodiversity, climate change, desertification, endangered species, law of the sea, nuclear test ban, ozone layer protection.

Appendix 3. Basic Statistics and Population

Physical areas			
Area of country	2002	250581000	ha
Cultivated area (arable land and area under permanent crops)	2002	16653000	ha
as % of the total area of the country	2002	7	%
Arable land (annual crops + temp. fallow+temp. meadows)	2002	16233000	ha
Area under permanent crops	2002	420000	ha
Population			
Total population	2004	34333000	inhabitants
Of which rural	2004	60	%
Population density	2004	14	Inhabitants/km^2
Economically active population	2004	13806000	inhabitants
As % of total population	2004	40	%
Female	2004	30	%
Male	2004	70	%
Population economically active in agriculture	2004	7925000	inhabitants
As % of total economically active population	2004	57	%
Female	2004	38	%
Male	2004	62	%
Economy			
Gross domestic product (GDP)	2003	17800	Million US$/yr
Value added in agriculture % of (GDP)	2002	39.2	%
GDP per capita	2003	518	US$/yr
Human development index (highest=1)	2002	0.505	
Access to improved drinking water sources			
Total population	2002	69	%
Urban population	2002	78	%
Rural population	2002	64	%

Appendix 4. Haffirs

The Haffir is the local name in Sudan for water reservoir. The Haffir is a hollow dug in the ground designed to store water runoff after a rainy season; the Haffir is usually used in semi-arid regions where rainfall is annual but over short periods and storage is required for the rest of the year. The Haffir can be natural or manmade; water storage is not a new concept, however, the technology of today can improve the efficiency of the traditional water-storing methods. The water is used by all the community, farmers, nomads, livestock and for domestic drinking water.

Requirements

- Situated in clay soils so filtration is reduced, allowing maximum storage and less labor requirements to provide an impermeable base.
- Near an annual water source, allowing the Haffir to be refilled, as well as filling from surface runoff and its own catchments.
- Basic filtration system, to prevent the deposition of silt in the Haffir.

Layout

The Haffir is often a considerable distance from the area that it is supplying. This is because of the requirement for a clay soil type and a nearby annual water source; a Haffir is normally found in areas where no underground occurs. This formation is known as a "basement complex;" because of this, many towns in Sudan are reliant on the water that haffirs can store. Water from the source is diverted to the Haffir through a dug canal. The Haffir can vary in size and shape; an average volume would be 30,000 m^3. The Haffir is often conical in shape with an average depth of approximately 3.5 m.

The Azagarfa Haffir in Darfur, Sudan.

INDEX

A

B

C

D

E

F

G

T

U

V

W